Text Book of

MICROWAVE

AF897610

(ELECTIVE)

For
Third Year Diploma – Semester VI
Diploma in Electronics Engineering Groups
(ECE 608)

As Per New Syllabus of SBTE, Jharkhand

Vijay G. Yangalwar
M.Sc. LL.B., L.M.I.S.T.E.
Retd. HOD, Deptt. of Electronics Engineering,
Government Polytechnic, Sadar, Nagpur.

NIRALI PRAKASHAN
ADVANCEMENT OF KNOWLEDGE

N4750

Microwave (Sem. VI) **ISBN 978-93-89944-33-4**

First Edition : **February 2020**

Published By :
NIRALI PRAKASHAN
Abhyudaya Pragati, 1312, Shivaji Nagar
Off J.M. Road, PUNE – 411005
Tel - (020) 25512336/37/39, Fax - (020) 25511379
Email : niralipune@pragationline.com

➢ DISTRIBUTION CENTRES

PUNE

Nirali Prakashan (For orders within Pune) : 119, Budhwar Peth, Jogeshwari Mandir Lane, Pune 411002, Maharashtra
Tel : (020) 2445 2044, Fax : (020) 2445 1538; Mobile : 9657703145
Email : niralilocal@pragationline.com

Nirali Prakashan (For orders outside Pune) : S. No. 28/27, Dhayari, Near Asian College Pune 411041
Tel : (020) 24690204 Fax : (020) 24690316; Mobile : 9657703143
Email : bookorder@pragationline.com

MUMBAI

Nirali Prakashan : 385, S.V.P. Road, Rasdhara Co-op. Hsg. Society Ltd.,
Girgaum, Mumbai 400004, Maharashtra; Mobile : 9320129587
Tel : (022) 2385 6339 / 2386 9976, Fax : (022) 2386 9976
Email : niralimumbai@pragationline.com

➢ DISTRIBUTION BRANCHES

JALGAON

Nirali Prakashan : 34, V. V. Golani Market, Navi Peth, Jalgaon 425001, Maharashtra,
Tel : (0257) 222 0395, Mob : 94234 91860; Email : niralijalgaon@pragationline.com

KOLHAPUR

Nirali Prakashan : New Mahadvar Road, Kedar Plaza, 1st Floor Opp. IDBI Bank, Kolhapur 416 012
Maharashtra. Mob : 9850046155; Email : niralikolhapur@pragationline.com

NAGPUR

Nirali Prakashan : Above Maratha Mandir, Shop No. 3, First Floor,
Rani Jhanshi Square, Sitabuldi, Nagpur 440012, Maharashtra
Tel : (0712) 254 7129; Email : pratibhabookdistributors@gmail.com

DELHI

Nirali Prakashan : 4593/15, Basement, Agarwal Lane, Ansari Road, Daryaganj
Near Times of India Building, New Delhi 110002 Mob : 08505972553
Email : niralidelhi@pragationline.com

BENGALURU

Nirali Prakashan : Maitri Ground Floor, Jaya Apartments, No. 99, 6th Cross, 6th Main,
Malleswaram, Bengaluru 560003, Karnataka; Mob : 9449043034
Email: niralibangalore@pragationline.com

Other Branches : Hyderabad, Chennai

Preface ...

I am glad to present the book entitled **"Microwave"** for **Third Year (Sixth Semester) Diploma in Electronics Engineering** as per SBTE's New Revised syllabus.

I have observed the students facing extreme difficulties in understanding the basic principles and fundamental concepts. To meet this basic requirement of students, sincere efforts have been made to present the subject matter with frequent use of figures.

I am thankful to Publisher Mr. Dineshbhai Furia, Mr. Pradeepbhai Furia, Mr. Jigneshbhai Furia and Staff of Nirali Prakashan specially Mr. P. M. More, Mr. Kiran Velankar, Mr. Ilyas Shaikh, Mrs. Anjali Muley, Mr. Ravindra Walodare for bringing this publication timely to the Market.

I specially thank my students for motivating and encouraging us from time to time.

Any errors or suggestions for the improvement of this book brought to notice will be thankfully acknowledged and incorporated in the next edition.

Vijay G. Yangalwar

Syllabus ...

Chapter 1 : Microwave Tubes **(10 Hours)**

Limitations of Conventional Tubes, Construction, Operation and Properties of Klystron Amplifier, Reflex Klystron, Magnetron, Travelling Wave Tube (TWT), Backward Wave Oscillator (BWO), Crossed Field Amplifiers.

Chapter 2 : Microwave Solid State Devices **(11 Hours)**

Limitations of Conventional solid state devices at Microwaves, Transistors (Bipolar, FET), Diodes (Tunnel, Varactor, PIN), Transferred Electron Devices (Gunn diode), Avalanche transit time effect (IMPATT, TRAPATT, SBD), Microwave Amplification by Stimulated Emission of Radiation (MASER).

Chapter 3 : Microwave Components **(11 Hours)**

Analysis of Microwave components using s-parameters, Junctions (E, H, Hybrid), Directional coupler, Bends and Corners, Microwave posts, S.S. tuners, Attenuators, Phase shifter, Ferrite devices (Isolator, Circulator, Gyrator), Cavity resonator, Matched termination.

Chapter 4 : Microwave Measurements **(10 Hours)**

Power measurements using Calorimeters and Bolometers, Measurement of Standing Wave Radio (SWR), Frequency and wavelength, Microwave bridges.

✍✍✍

Contents ...

✍✍✍

Chapter 1

MICROWAVE TUBES

Syllabus

Limitations of Conventional Tubes, Construction, Operation and Properties of Klystron Amplifier, Reflex Klystron, Magnetron, Travelling Wave Tube (TWT), Backward Wave Oscillator (BWO), Crossed Field Amplifiers.

1.1 FUNDAMENTALS OF MICROWAVE

1.1.1 Introduction

- During World War II, microwave engineering became a very essential consideration for the development of high resolution Radars capable of detecting and locating energy planes and ships through a narrow beam of Electromagnetic (EM) energy. Such a beam could be achieved by a *paraboloid antenna* of large diameter compared to the wavelength of radiation.
- Hence, the size of the antenna can be carried by an airplane or board can be reduced by using such wavelength of the *microwave*. The microwave band is capable of making economic transmission of a large number of communication channels or TV programs by modulating these channels into a single microwave carrier and transmitting them over one few percent of the microwave carrier frequency.
- Because of very short wavelengths, microwaves are capable of freely propagating through the ionized layers in the atmosphere. So, the application in radio astronomic research of space and communication between the ground stations and space vehicles.
- In recent years, microwaves are widely used in domestic micro ovens for rapid cooking, and also in industrial and medical uses for microwave heating.
- Microwave transmission involves propagation of *electromagnetic (EM) waves* consisting of changing electric and magnetic fields in the medium. The specific properties of microwave develop techniques for microwave-guided structures, such as waveguides, co-axial lines, strip lines, microstrip lines, slot lines and microwave sensors like magnetrons, travelling wave tubes, backward wave, oscillators, klystrons, etc.

1.1.2 Definition

- **Microwave is a region in the electromagnetic (EM) spectrum in the frequency range from 0.3 GHz to 100 GHz or even extended upto 300 GHz.** This corresponds to a range from 30 cm to 0.3 mm in free space.

1.1.3 Microwave Frequency Spectrum

- The electromagnetic (EM) wave spectrum region allotted is from 0.3 GHz to 100 GHz and extended upto 300 GHz.
- **The entire range of frequency of microwave region from 0.3 GHz to 100 GHz is known as microwave spectrum.**

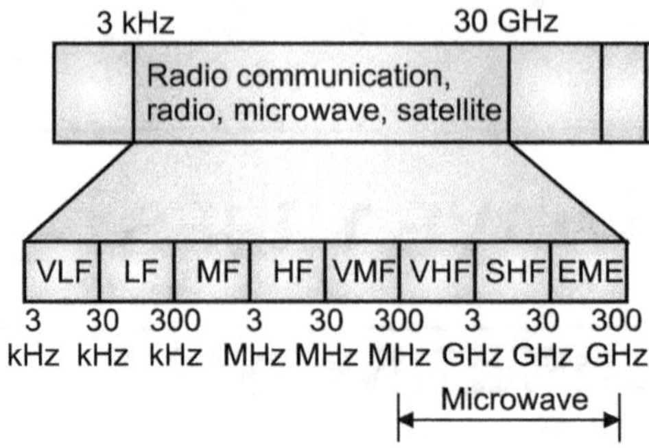

- VLF - Very low frequency
- LF - Low frequency
- MF - Medium frequency
- HF - High frequency
- VHF - Very high frequency
- UHF - Ultra high frequency
- SHF - Super high frequency
- EHF - Extremely high frequency

Fig. 1.1 : Electromagnetic (EM) spectrum

- **The UHF, SHF and EHF regions of the electromagnetic (EM) frequency spectrum is known as microwave frequency signal.**
- The microwave frequency spectrum is as shown in Fig. 1.1.

1.1.4 Types of Microwave Band Regions

- The types of microwave band regions alongwith their frequency ranges are as shown in Table 1.1.

Table 1.1

Sr. No.	Band Name	Band Designation	Frequency Range
1.	UHF	Ultra High Frequency	300 MHz to 3 GHz
2.	SHF	Super High Frequency	3 GHz to 30 GHz
3.	EHF	Extra High Frequency	30 GHz to 300 GHz

1.1.5 Microwave Band Designations

Table 1.2

Sr. No.	Band Letter Designation	Frequency Range	Wavelength Range
1.	L Band	1 GHz to 2 GHz	15 cm to 30 cm
2.	S Band	2 GHz to 4 GHz	7.5 cm to 15 cm
3.	C Band	4 GHz to 8 GHz	3.75 cm to 7.5 cm
4.	X Band	8 GHz to 12 GHz	25 mm to 37.5 mm
5.	K_u Band	12 GHz to 18 GHz	16.7 mm to 25 mm
6.	K Band	18 GHz to 26.5 GHz	11.3 mm to 16.7 mm
7.	K_a Band	26.5 GHz to 40 GHz	5.0 mm to 11.3 mm
8.	Q Band	36 GHz to 50 GHz	6.0 mm to 9.0 mm
9.	U Band	40 GHz to 60 GHz	5.6 mm to 7.5 mm
10.	V Band	50 GHz to 75 GHz	4.0 mm to 6.0 mm
11.	W Band	75 GHz to 110 GHz	2.7 mm to 4.0 mm
12.	F Band	90 GHz to 140 GHz	2.1 mm to 3.3 mm
13.	D Band	110 GHz to 170 GHz	1.8 mm to 2.7 mm

1.1.6 Field of Applications

- The field or areas of microwaves are as under :

1. Communication,
2. Spectroscopy,
3. Power,
4. Remote sensing.

1.1.7 Advantages of Microwave

1. It has increased bandwidth availability.
2. It has improved directive properties.
3. There is less fading effect at high frequencies and hence, it is more reliable.
4. It has very low power requirements.
5. It makes it possible for duplex communication and exchange of information between ground stations and space vehicles.

1.1.8 Applications

1. **Communication :**
   - Long Distance Telephone Communication.
   - Mobile Phones Two-Way Radio Communications.
   - Amateur Radio Satellite Communication.
   - Space Communications, Terrestrial Communications.
   - Wireless Internet Accessories.
   - Metropolitan Area Network (MAN).
   - Broadband Wireless.
   - Wide Area Mobile Broadcasting (WAMB).
   - GSM.
   - Wireless LAN.
   - Bluetooth.
   - Military Communication.
   - GPS.
   - Military Telemetry.
   - Satellite TV Broadcasting, etc.
2. **Spectroscopy :**
   - Electronic Paramagnetic Resonance Spectroscopy.
3. **Power :**
   - Microwave ovens.
   - Industrial processes for drying and curing products.
   - Microwave heating.
   - Generation of plasma for ion etching chemical vapour deposition.
   - Solar panel satellite etc.
4. **Remote Sensing :**
   - Remote range, speed and other characteristics of remote objects.
   - Air Traffic Control.
   - Weather Forecasting.
   - Navigation of Ships and Speed Limit Enforcement.
   - Automatic Door Openers.
   - Radio Astronomy.
   - Astronomical Observations.
5. **Radars :**
   - Weather Radar.
   - Surface Ship Radar.
   - Specified Radar.
   - Automotive Radar.
   - Military Wave Radar Research.
   - Scientific Research.
   - Radar Research.
   - Non-military Applications.
   - Most Modern Radars.
   - Directed Energy Weapons.
   - Millimeter Wave Scanner.
   - Military Radar for Targeting and Tracking Applications.

6. Microwave Devices and Components :

- Gunn diode
- IMPATT diode
- PIN diode
- Tunnel diode
- Varactor diode
- Step recovery diode
- Klystron tube
- Magnetron tube
- Travelling wave tube etc.

1.2 MICROWAVE TUBES

1.2.1 Introduction

- Like electromagnetic waves at lower frequency, microwaves can also be generated by using vacuum tubes or semiconductor devices. However, these devices and components are based on different principles, although in some cases (such as for transistors), frequency limit has been extended into the microwave range by the advancement of technology and innovations in the design.
- Vacuum tubes are still found in microwave equipments. This is true particularly in microwave transmitters used for producing high output power. Special microwave tubes developed during World War Second are still widely used for microwave power amplification. In satellite earth stations, TV stations, and in some military equipments such as radar, very high output powers are needed.

1.2.2 Limitations of Conventional Tubes

1. It suffers from inter-electrode capacitance effect.
2. It suffers from lead inductance effect, which limits its performance.
3. It suffers from transition time effect, which reduces the output.
4. It has limited gain - bandwidth product.
5. It suffers from skin effect losses and dielectric losses, which reduce its efficiency.
6. It has more radiation losses which reduce its efficiency. Due to this limitation, conventional tubes cannot be used for frequencies higher than 100 MHz.
7. It leads to stray reactances around 500 MHz.
8. The transit time becomes comparable to the time period of oscillations at microwave frequency.
9. The size reduces as the frequency is raised.

1.2.3 Need

- The conventional vacuum tubes suffer from the limitations of stray reactances, transit time and size. The basic limitation of nearly all the active devices at microwave frequencies is that of the size. As frequency is raised, obviously, the size of the conventional devices reduces. Consequently, the powers handling fall and noise levels rise. The net effect at microwave frequencies is that the conventional lower frequency vacuum tubes fail to generate large power. Therefore, there is a need to design the special types of vacuum tubes, i.e. microwave tubes to generate microwave power.

1.2.4 Classification of Linear Beam Tubes

The classification of linear beam tubes is as shown in Fig. 1.2.

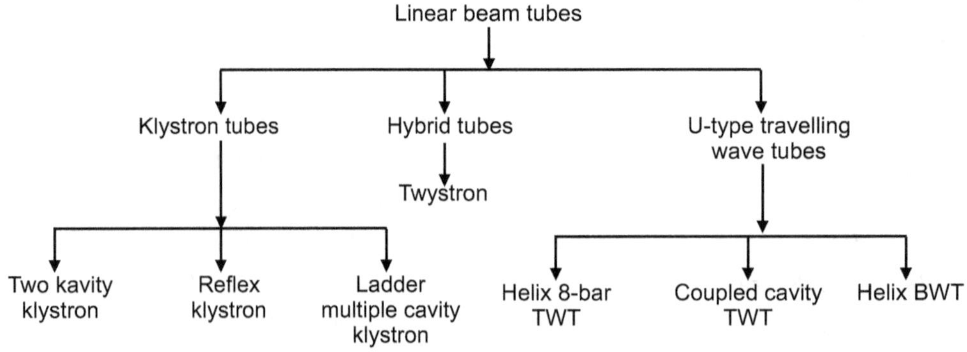

Fig. 1.2 : Classification of linear beam tubes

1.3 KLYSTRON AMPLIFIER TUBES

1.3.1 Introduction

- A Klystron was the first significantly powerful source of radio waves in the microwave range. It was invented in 1937 by two American electrical engineer brothers *Russel* and *Sigural Varian* at Stanford University as a source and amplifier of microwave.
- It provides much high powers that previously been obtainable at microwave frequencies, but at very low powers.

1.3.2 Definition

- **Klystron is a linear beam microwave vacuum tube using cavity resonators to produce velocity modulation of the electron beam and to produce amplification.**

1.3.3 Types

Klystron amplifiers are of two types :

1. Multicavity Klystron.
2. Reflex Klystron.

1.4 TWO-CAVITY KLYSTRON AMPLIFIER

1.4.1 Cavity Resonator

- Let us consider the structure of a cavity resonator as shown in Fig. 1.3.
- An electron gun is used to produce a high velocity narrow beam of electrons that passes through the fields in a resonant cavity.
- The electron gun consists of a cathode C, and tungsten filament F used to heat this cathode.
- The anode is kept at a positive potential and accelerates the electrons emitted by the cathode.

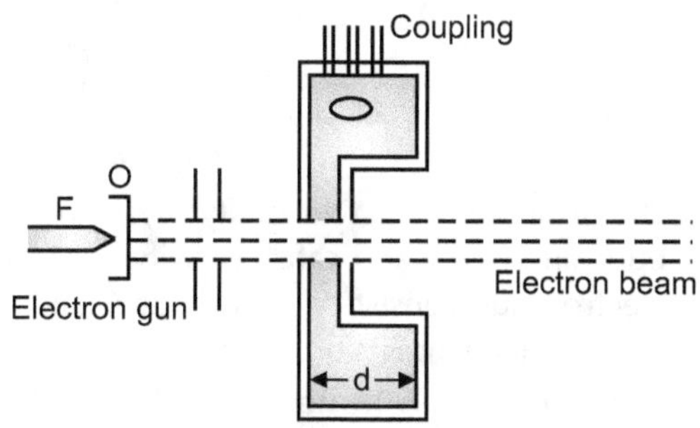

Fig. 1.3 : Cavity resonator

- It has an opening in the middle for the accelerated beam of electrons to pass through. In addition, there are one or more suitably shaped electrodes, which help to focus the beam. This electron beam is later collected by a collector electrode called *anode*.
- The electromagnetic power can be coupled into or out of the cavity through a coupling loop arrangement shown in Fig. 1.3. The electric field component in the cavity is collinear with the beam axis. Thus, the electrons passing through the resonator see an accelerating field for half a cycle and a retarding field for the other half time period. Consequently, some of the electrons will be accelerated and some retarded. This phenomenon is called the *velocity modulation* of the beam.
- As the electrons move ahead, at the average velocity of the beam, the velocity modulation causes a bunching of electrons. The bunching effect converts the velocity modulation into the density modulation of the beam. This may be illustrated with the help of a distance-time diagram. The representation of distance-time diagram is known as Applegate diagram. As the bunches of electron pass through the catcher cavity, the cavity is excited into oscillations at radio frequencies (RF). The d.c. kinetic energy in the electron beam is converted into RF energy at the cavity frequency, thus application takes place.

1.4.2 Velocity Modulation and Bunching Processes

- Let us consider the interaction between an electron beam and an electromagnetic field of a cavity resonator as shown in Fig. 1.3.
- The electromagnetic power can be coupled into or out of the cavity through a coupling loop arrangement as shown in Fig. 1.3.
- The electric field component in the cavity is collinear with the beam axis.

- The electron passing through the resonator results in accelerating field for half a cycle and retarding field for other half time period. This phenomenon is called as *velocity modulation*. At the cavity distance d, the density of electrons in the electron beam varies periodically with time.
- Under these conditions, the electron beam contains RF component of current and is said to density modulated.
- Thus, as a result of the drift of the electron beam over a distance 'd' cavity, the velocity modulation is converted into density modulation.
- As the bunches of electrons pass through the catcher cavity, the cavity is excited into oscillations at RF. The d.c. kinetic energy in the electron beam is converted into RF energy at the cavity frequency.

1.4.3 Construction

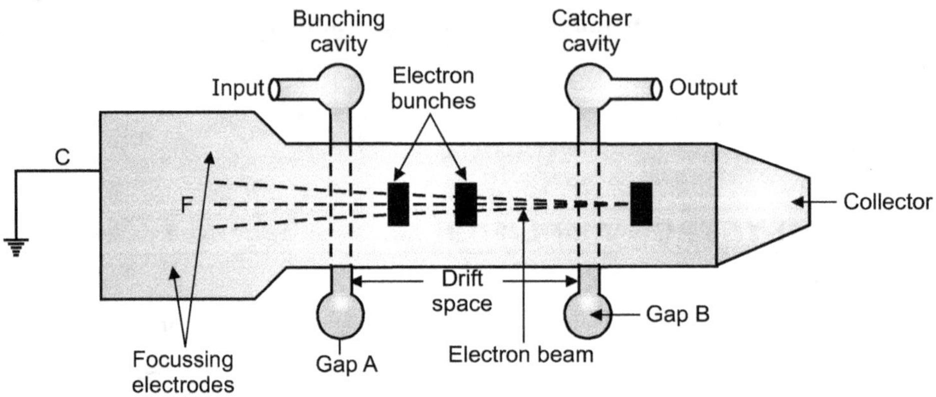

Fig. 1.4 : Two cavity Klystron amplifier

- The structure of a two cavity Klystron amplifier is as shown in Fig. 1.4. It consists of an electron gun, bunching cavity and catcher cavity as shown in Fig. 1.4. Both the cavities are open in the center area. The electron gun consists of a cathode C and filament F. The cathode C is usually made of oxide coated nickel and the filament F is usually made of tungsten.
- The cathode C is heated by a filament F. At very high temperature, the cathode C emits electrons. The electrons emitted by the cathode are focused into a very narrow stream by electrostatic and electromagnetic focusing techniques. The sharply focused beam of electrons is then forced to pass through the centers of the two cavity resonators.

1.4.4 Concept of Velocity Modulation

- It is assumed that the oscillations set-up in the tube initially due to noise or switching transients and these oscillations are sustained by device operation. This can be explained by Applegate diagram as shown in Fig. 1.5.
- The RF voltage that is produced across the gap by the cavity oscillations act on the electron beam to cause velocity modulation.
- 'e_R' is the reference electron that passes through the gap voltage which is zero and going negative.
- Electrical 'e_R' is unaffected by the gap voltage. This moves towards the repeller and gets reflected by the negative voltage on the repeller.
- It returns and passes through the gap for a second time.

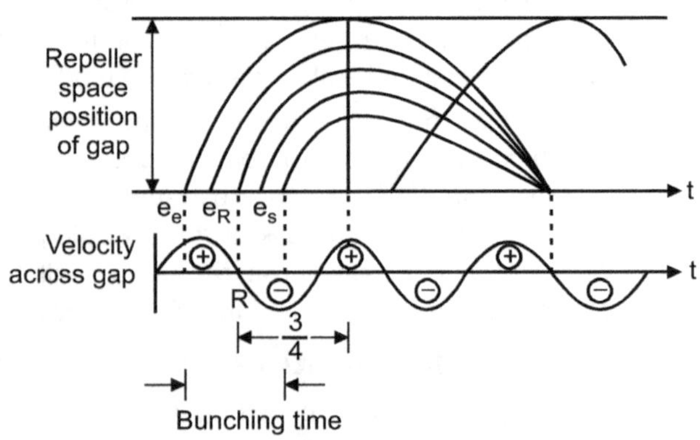

Fig. 1.5 : Applegate diagram

- The early electron 'e_e' that passes through the gap before the reference electron 'e_R' experiences a maximum positive voltage across the gap and this electron is accelerated. It moves with greater velocity and penetrates depth of penetration into the repeller space is more. Hence, 'e_e' and 'e_R' operate at the gap for the second time at the same time. The last electron 'e_p' that passes the gap later than reference electron 'e_R' experiences a maximum negative voltage and moves with a retarding velocity.

- The return time is shorter as penetration into repeller space is less and catches up will e_R and e_e electrons forming *electron bunch*. Bunches occur once per cycle centred around the reference electron 'e_R' and these bunches transfer maximum energy to get sustained oscillations.
- For oscillations to be sustained, the time taken by electrons to travel into the repeller space and back to the gap called as *transit time* must have an optimum value. The factor is not important in a klystron amplifier, but it assumes a great importance here. In general, the optimum transit timer should be,

$$\tau = n + \frac{3}{4} \qquad \text{... (1.1)}$$

where, n = Any integer

- The integer depends on repeller and anode voltage.

1.4.5 Principle of Operation

- Consider Fig. 1.6 showing the structure of a two cavity Klystron amplifier. An electron gun produces a high velocity narrow beam of electrons that passes through the fields in a resonant cavity. This electron beam is later collected by a collector. A electrode which is kept at highly positive potential is called an anode.
- The shape of the cavity structure is cylindrical and is symmetrical about the beam axis. The microwave signal to be amplified is applied to the lower cavity through the coupling loop. This sets up electric and magnetic fields in the cavity, which influence the electron beam. As a result of this, they cause the electrons to speed up and slow down as they pass through the cavity. Thus, on one half cycle, the electrons are speeded up and on the next half cycle of the input, they are slowed down.
- The effect of this is to create punches of electron that create bunches of electrons that are one-half wavelength apart in the drift space between the cavities. This speeding up and slowing down or the electron beam is called *velocity modulation*.
- These bunched electrons move towards the collector in clouds of the alternately dense and sparse areas. So, we can say that the electron beam is density modulated. As the bunches of electrons pass through the catcher cavity, the cavity is excited into oscillations at the resonant frequency.
- Therefore, the d.c. kinetic energy in the electron beam is converted into the RF energy at the cavity frequency. Thus, the amplification tubes place. The RF output is taken from the C other cavity.

1.4.6 Applegate Diagram

- As the electron move ahead, at the average velocity of the beam, the velocity modulation causes a bunching of the electrons. The bunching effect converts the velocity modulation into the density modulation of the beam.
- This may be illustrated with the help of a distance time diagram as shown in Fig. 1.6.
- The graphical representation of electrons by a distance-time graph is known as Applegate diagram.
- Electrons that enter during the time interval A to B move slower and these enter between B and C move faster.
- As a result of these differences in velocities, electrons tend to cluster around a point D at a distance 'd'.

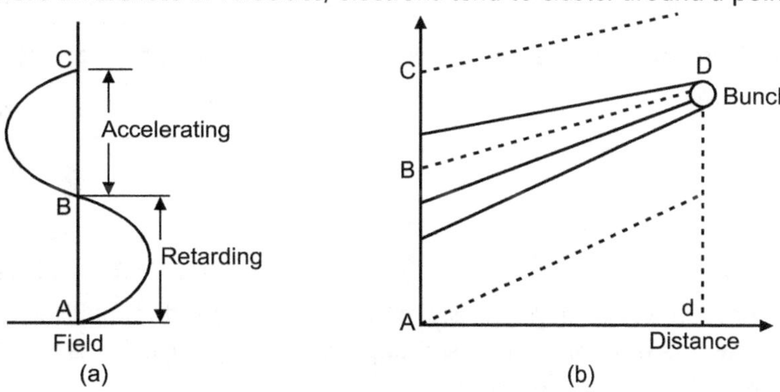

Fig. 1.6 : Applegate diagram

1.4.7 Applications

1. It is used to obtain medium, high or very high power amplifiers in the UHF and microwave ranges.
2. It is used in satellite station transmitter as RF power amplifier.
3. It is used in UHF TV transmitter.
4. It is used long range radars.
5. It is used in liner practical accelerator.
6. It is used in earth station transmitter.

1.4.8 Disadvantages

1. It suffers from noise caused due to bunching.
2. This makes them to noisy for use in receivers and transmitters.

1.5 REFLEX KLYSTRON

1.5.1 Introduction

- To use the Klystron as an oscillator, we require a variable phase shifter between two cavities to assure a positive feedback at the desired frequency of oscillation.
- This scheme is seldom used. A Klystron as an oscillator which can be designed by using only a single resonant cavity. This is known as reflex Klystron oscillator. It provides low power and has low efficiency.

1.5.2 Definition

- **A reflex Klystron is a low power, low efficiency microwave oscillator (or source) that works on reflections and oscillations in a single resonant cavity which has a variable frequency.**
- Reflex Klystron has only one resonant cavity through which electrons pass twice.
- Reflex Klystron uses only a single cavity resonator and operates as an oscillator.

1.5.3 Salient Features

1. It is a small and low power device.
2. It generates power above 100 MW.
3. It generates oscillations for frequencies ranging from 4 GHz to 20 GHz.

1.5.4 Basic Principle

- The electrons pass through the cavity and undergo velocity modulation, which produces electron bunching. The distance between the repeller and cavity is such that cavity at the proper, time be in synchronization, because of this they deliver RF energy to the cavity that results in oscillations at the cavity frequency.

1.5.5 Construction

- Fig. 1.7 shows a schematic diagram of a reflex klystron.
- The various parts of a reflex klystron are as under :
  1. Electron gun,
  2. Cathode filament,
  3. Anode cavity,
  4. Cavity resonator,
  5. Repeller,
  6. Output coupling.

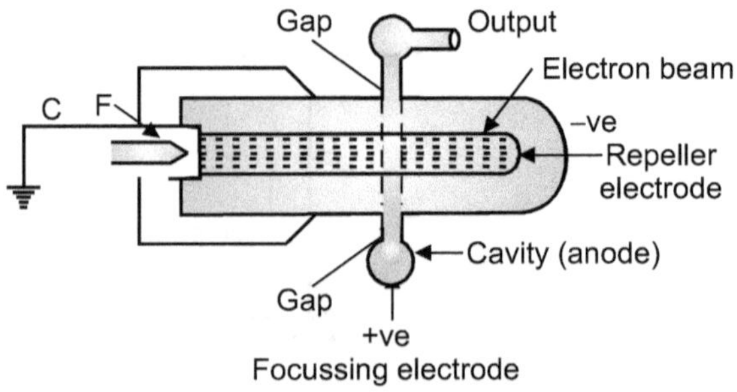

Fig. 1.7 : Reflex Klystron oscillator

- The electron gun used in a reflex klystron is similar to the one used in multicavity Klystron, but is much smaller comparatively. The electron gun is used to produce a high velocity narrow beam of electrons that passes through the fields in a resonator cavity.

- The magnetic focusing is not used, because this device is very small. The cavity acts as an anode and it has a high positive voltage. The repeller electrode is at a negative potential and sends the partially bunched electron beam back to the resonant cavity. This provides a positive feedback mechanism which support oscillations.

1.5.6 Principle of Operation

- Consider a reflex Klystron as shown in Fig. 1.7. The cathode C emits electrons, which are accelerated forward by an accelerating grid with a positive potential on it and focused into a narrow beam of electrons. These electrons pass through the cavity and undergo velocity modulation, which produces the electron bunching.
- The electron bunches move towards the repeller plate, which is made negative with respect to cathode. This causes the electrons to be repelled rather than attracted as they are by a positive collector in a standard Klystron. The result is that the bunched electrons are forced back through the cavity.
- The distance between the repeller and the cavity is so chosen that the cavity at proper time to be in synchronization. Because of this they deliver RF energy to the cavity. The result is oscillation at cavity frequency.
- The cavity itself is made positive, so that the electrons are ultimately attracted by the cavity and cause direct current (d.c.) flow in the external circuit. A coupling loop in the cavity removes the RF energy.

1.5.7 Bunching Process

- Let us consider an electron A, which passes through the resonator gap, when the voltage across the gap is zero. This electron returns to resonator at a time t_0.
- Later another electron B passes through the gap Δt earlier.
- Since gap voltage is positive at this initial velocity V_0 for this electron is higher than the average value than that of electron A.
- Thus, electron B would travel further towards the repeller and take a longer time to come back to the resonator.
- It would thus tend to bunch with electron A. Similarly, an electron C that passes through the gap Δt, after A, appears with a smaller initial velocity and so takes less time to come back.

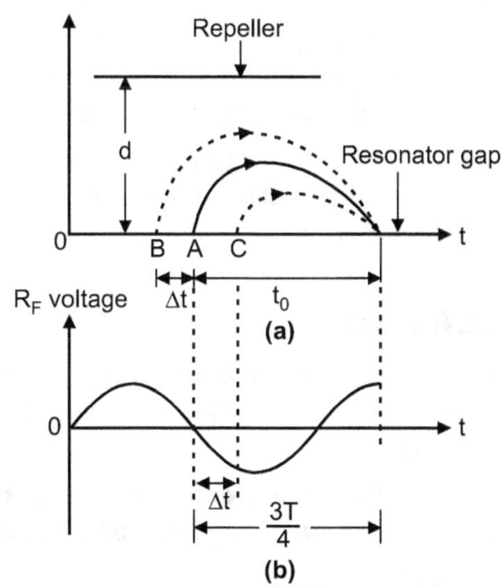

Fig. 1.8 : Parabolic trajectory of an electron

- Thus, electron C catches up with electron A. This gives rise to a 'bunch' around the electron A.
- The electron bunches so formed would deliver the RF power to the resonator, if they pass through the resonator at an instant, when the field retards the bunch.
- The parabolic trajectories of electron in the region between the resonator and the repeller is shown in Fig. 1.8. We note that this happens, when $t_0 = \frac{3T}{4}$, where T is the time period at the resonance frequency.

1.5.8 Applications

1. It is used as a signal source of microwave generator.
2. It is used as a local oscillator of microwave receivers.
3. It is used as a frequency modulated oscillator in portable microwave links.
4. It is used as a pump oscillator for parametric amplifiers.

5. It is used as a stable oscillator in radar receivers.
6. It is used as a submillimeter and millimeter oscillator.

1.6 CAVITY MAGNETRON

1.6.1 Brief History

- In 1910, *Hans Gordian* of the Siemens Corporation invented a magnetron.
- In 1912, Swiss physicist *Heinrich Greinacher* was looking for new ways to calculate the electron mass. But he failed to measure the electron mass, because he was unable to achieve a good vacuum in the tube.
- However, as a part of this work, he developed mathematical models of the electrons in the crossed electric and magnetic fields.
- Another form of magnetron was invented by *Albert Hull* of General Electric Research Laboratory in 1920, but it achieved only at a frequency of 30 kHz.
- An important advance was the multi-cavity magnetron, first proposed in 1934 by A. L. Samuel of Bell Telephone Laboratories.
- However, the first successful example was developed by *Alaksereff and* Molearoff in USSR in 1936, which achieved 300 watts at 3 GHz, i.e. 10 cm wavelength.
- The cavity magnetron was radically improved by *John Randall* and *Harry Boot* in 1940 at the University of Birmingham, England.

1.6.2 Introduction

- Electrons pass by the openings to these cavities and cause radio waves to oscillate within, similar to the way a whistle produce a tone, when excited by an air stream blown past its opening. The frequency microwave produced by the resonant frequency, is determined by the cavities of physical dimensions.
- Unlike other vacuum tubes such as a Klystron or a travelling wave tube (TWT), the magnetron cannot function as an amplifier in order to increase the intensity of an applied microwave signal. The magnetron serves solely as an oscillator, generating a microwave signal from direct current electricity supplied to the vacuum tube.

1.6.3 Definition

- The cavity magnetron is a high powered vacuum tube that generates microwave using the interaction of a stream of electrons with a magnetic field, while moving past a series of open metal cavities, i.e. cavity resonator.
- **A magnetron is a diode, which uses the interaction of magnetic and electric fields in a complex cavity to provide oscillations of very high peak powers above 100 kW and 3 GHz.**

1.6.4 Construction

- The cross-section of a hole and slot in magnetron tube is shown in Fig. 1.9. A magnetron usually has a cylindrical construction.
- The most common RF structure of a magnetron consists of a number of cylindrical cavities opening in an annular space called interaction region. Its anode structure has a number of permanent cavities as shown in Fig. 1.9.
- The cylindrical cathode is surrounded on all sides by a number of cavities cut in the anode block. The diameter of each cavity is equal to one-half wavelength of the desired operating frequency. The anode is usually made up of copper and is connected to a high voltage positive d.c. source.
- It uses an axial magnetic field and a radial electric field. The outer anode results in a d.c. radial electric field and the axial magnetic field is obtained by a magnet. The magnetic lines of force pass through the cathode and also through the interaction space. These lines of force are thus at right angles to the structure cross-section. The output can be taken out from any one of the cavities by means of a coaxial line or a waveguide, depending upon the frequency and power.

1.6.5 Principle of Operation

- Consider the magnetron as shown in Fig. 1.9. The cylindrical cathode emits the electrons, when heated. These electrons flow directly from the cathode straight to the anode, causing high current to flow.
- This direction of electrons is modified by the strong magnetic field supplied by a C-shaped permanent magnet. As the electrons come across the electric field in the direction of its velocity, the magnetic field interact and the path of electrons will be curved. By properly adjusting the anode voltage and the strength of the magnetic field, the electrons can be bend such that they rarely reach the anode. The path of electrons becomes circular. This causes the current flow.
- As a result of this, the electron is retarded by the field, slowed down and drifts towards the anode with a circular direction which causes the current to flow. In making their passes in the interaction chamber, the electrons exit the resonant cavities into oscillations. Therefore, the magnetron is an oscillator, not an amplifier.
- Such electrons extract their RF energy from the field. However, as these electrons quickly get charged from the interaction region, they absorb much less power than delivered by the favourable electrons. The RF power may be coupled out by a coaxial loop by means of an iris coupling the cavity to a waveguide outside.

1.6.6 Effects of Magnetic Field

- When an electron moves, it constitutes an electric current. If a magnetic field is applied, on the moving electron, a force is exerted on the electron, whose magnitude is proportional to the product BVe, where V and e are velocity and charge of the electron respectively. B is the magnetic field component in a plane perpendicular to the direction of electron motion.
- The field exerted on the electron is at right angles to the two directions. As the magnetic field is constant, its force on the electron depends solely on the forward (radial) velocity of the electron.

1.6.7 Effects of Electric Field

- The electric field simultaneously act along with the magnetic field on the electron. This results in a change in path of the electron and the shape of this path is governed by the relative strengths of the mutually perpendicular electric and magnetic fields.
- Fig. 1.9 shows some of these electron paths in the absence of oscillations in a magnetron, in which the electric field is constant and radial, and the axial magnetic field can have any value.
- When the magnetic field is zero, the electron goes straight from the cathode to the anode, accelerating all the time under the influence of the radial electric field, indicated by the path 1.

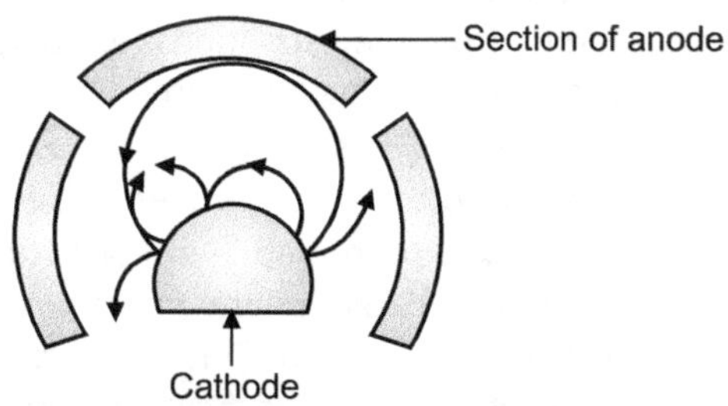

Fig. 1.9 : Electron paths

- When the magnetic field is small, it exerts a lateral force on the electron, making it to bend its path to the left, as shown by path 2. As the electron approaches the anode, its velocity continues to increase radially due to the accelerating force of the electric field. This makes the effect of magnetic field more pronounced, so that the path of curvature becomes more and more circular as the electrons approach the anode.

1.6.8 Phase Focussing Effect

- An electron 'b' is accelerated by the RF field and instead of imparting energy to the oscillations take energy from oscillations, resulting in increased velocity. Hence, bends more sharply, spends very little time in the interaction space and is *returned back to the cathode.*
- Such electrons are called *unfavoured electrons*. These electrons do not participate in bunching process rather they cause back heating.

- Similarly, an electron 'a' which is emitted a little later to be in correct position moves faster and tries to catch up with electron 'a' and an electron emitted at 'd' will be slowed down to fall back in step with electron 'a'.
- This results in all favoured electrons, like a, b, c, d to form a bunch one for each two modes as shown in Fig. 1.10.
- This process is called phase focussing effect and corresponds to reference electrons 'a'.
- The phase focussing effect of these favoured electrons imparts enough energy to the oscillations, so that they are sustained.

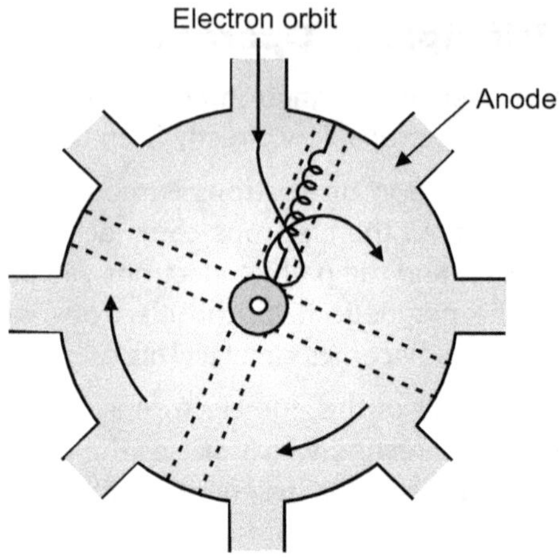

Fig. 1.10 : Phase focussing effect

1.6.9 Sustained Oscillations in Cavity Magnetron

- Assume that the RF oscillations are initiated due to some noise transient within the magnetron and oscillations are sustained by the device operation. When n = 4, then there is π-mode of operation, which is shown in Fig. 1.11. The anode poles are π-radiation apart in phase.
- As shown in Fig. 1.11, the electron 'a' is seen to be slow down in the presence of oscillations, thus, transferring energy from cathode to anode.

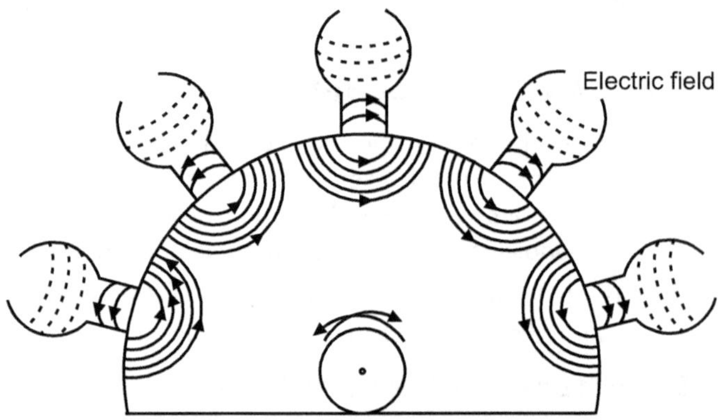

Fig. 1.11 : π-Mode of magnetron

- The electrons which participate in transferring the energy to the RF field are called *forward electrons* and they are responsible for the bunching effect. The electron 'b' is accelerated by the RF field and it takes energy from the oscillations resulting in increased velocity. It bends more sharply, spends very little time in interaction space and returned back to the cathode.
- The electron 'd' slowed down and it fills back in step with electron 'a'. This results in forward electron like a, b, c, d to confine to spokes of electron clouds.
- The spokes so formed in π-mode rotate with an angular velocity corresponding to two poles per cycle. The process is called *phase focussing effect* corresponding to a bunch of forward electrons around the reference electron 'a'. This phase focussing effect of these forward electrons imparts enough energy to the RF oscillations so that they are sustained.

1.6.10 Specifications

1. The maximum available power ranges from 10 MW in the UHF band.
2. Voltage Tunable Magnetrons (VTMs) available for frequency ranges from 200 MHz to X band CW powers upto 1 kW.
3. For fixed frequency CW magnetron, operating frequencies are 900 MHz and 2.5 GHz, power ranges from 300 W to 10 kW.

1.6.11 Applications

1. It is used for generating pulse microwave power for Radars.
2. It is used in linear particle accelerators.
3. It is used for industrial heating and microwave ovens.
4. It is used as sweep oscillators in telemetry and in missile operation.
5. It is used in excited lighting systems, such as a sulphur lamp.

1.7 TRAVELLING WAVE TUBES (TWT)

1.7.1 Brief History

- The original design and prototype of the travelling wave tube (TWT) was done by *Andrey Andy Haeff C.* in 1931 while he was working as a doctoral student at the Kellage Rational Laboratory at Caltech.
- The invention of the TWT is often attributed to *Rudolf Kampfner* in 1942 - 1943.
- In addition, *Nils Lindenblad*, working at RCA (Radio Corporation of America) in the USA also filed a patent for a device in May 1940, that was remarkably similar to Kompfner's TWT.
- Kompfner developed his TWT in a British *Admiralty* Radar laboratory during World War II and he built his first TWT in early 1943. The TWT was later refined by Kompfner, Jon R. Pierce and Lester M. Field Laboratories.

1.7.2 Introduction

- In Klystrons and magnetrons, the microwave circuit consists of a resonant structure, which limits the operating frequency range, i.e. bandwidth. Travelling Wave Tube (TWT) was invented by a British Research *Rudolf Kompfner* at the University of Birmingham in Britain and it was presented in the United States after World War II. Like a multicavity Klystron, it is a linear-beam tube used as a microwave amplifier. Unlike a Klystron, it is a device, in which the interaction between the electron beam and the RF field is continuous.

1.7.3 Definition

- **A travelling wave tube (TWT) is a broadband microwave active device.**
- It has extremely wide bandwidth of operation. It has ability to generate hundred and thousand watts of microwave power. It is not resonant at a single frequency.
- A travelling wave tube (TWT) is a specialized vacuum tube that is used in electronics to amplify RF signals in the microwave range. The TWT belongs to a category of *linear beam* tubes, such as the Klystron, in which the radio wave is amplified by absorbing from a beam of electrons as it passes down the tube.

1.7.4 Types

- There are various types of TWT, two major categories are as under :
  1. Helix TWT,
  2. Coupled cavity TWT.

1.7.5 Basic Principle

- In travelling wave tube, the interaction between the electron beam and the RF field is continuous. So, the RF wave on helix grows and reaches it to maximum at the output end. Therefore, it is important that both of them move in the same direction with nearly the same speed.
- Thus, to have a continuous interaction, a slow wave structure is used in TWT. In this, the RF field is retarded with the help of a helix or a waveguide coupled cavity arrangement.

1.7.6 Construction

- The construction of a typical travelling wave tube (TWT) is as shown in Fig. 1.12.

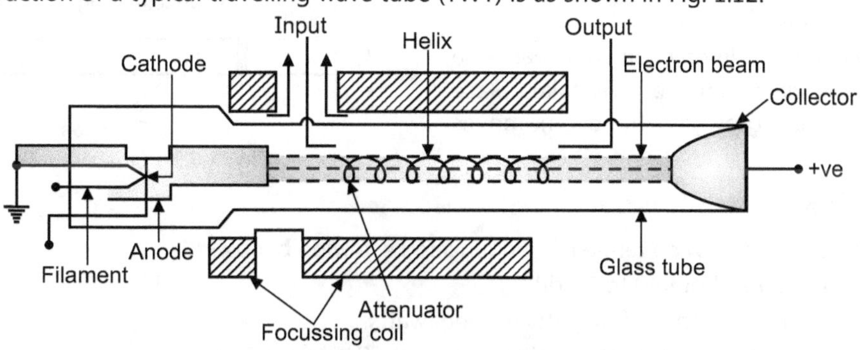

Fig. 1.12 : Helix type travelling wave tube

- It consists of two main constituents - an electron beam and a structure supporting a slow electro-magnetic wave. It has a pierce type electron gun similar to that used in Klystron amplifier.
- It has a coil in the form of helix, through which the electron beam is paused. Helix provides less change in phase velocity with frequency. Therefore, it is preferred over other slow-wave structure for TWT. Helix is a loosely wound thin conducting helical wire, which acts as a slow wave structure.
- Here the electron beam is required to travel a much longer distance and a focusing mechanism becomes necessary to prevent the electron beam from spreading. The electron beam is generated by an electron gun.
- The common method of beam focusing is by applying a longitudinal static magnetic field along the axis of the electron beam. This magnetic field has no effect on the electron travelling along or parallel to the beam axis. However, if an electron is diverging out and has a component of velocity in radial direction, the axial magnetic field produces a force on that electron.
- Thus, the electrons which try to diverge out due to space charge repulsion forces are bent into circumferential direction and start spiralling around the beam axis. This provides the required focussing mechanism.

1.7.7 Principle of Operation

- Consider a typical travelling wave tube as shown in Fig. 1.13. An electronic gun produces a very narrow beam of electrons, which is sent through the centre of a long axial helix. The helix is made positive with respect to the cathode and the collector.
- The electron beam is attracted to the collector and acquires high velocity. The narrow and correctly focussed electron beam passes through the centre of helix, without touching the helix itself. The signal to be amplified is applied at the input terminal of the helix through a waveguide.
- The RF field travels around the helix with a speed, which is nearly equal to the velocity of light in a free space. The axial speed of RF field is equal to the velocity of light multiplied by the ratio of helix pitch to helix circumference. This can be made quite low, approximately equal to the electron beam velocity.
- The axial RF field and the electron beam can now interact continuously with the electron beam bunching and giving the RF energy to the field.
- As a result, almost complete bunching takes place and so achieves high gain. The electrons leaving the cathode at random quickly encounter the work RF field at the input end of the helix, which is due to input signal. When the electromagnetic wave is travelling along the helix, there is a longitudinal component of RF electric field.
- This time varying field causes a velocity modulation in the electron beam passing through the helix axis.
- The velocity modulation tends to bunch the electrons at regular intervals of one wavelength as shown in Fig. 1.13.
- When the wave and the electrons are moving with equal velocities, the position of RF electric field and the bunches in the beam remain invariant with time.
- Thus a continuous interaction between the beam and the waves on the helix takes place and the bunches grow as the beam moves further.

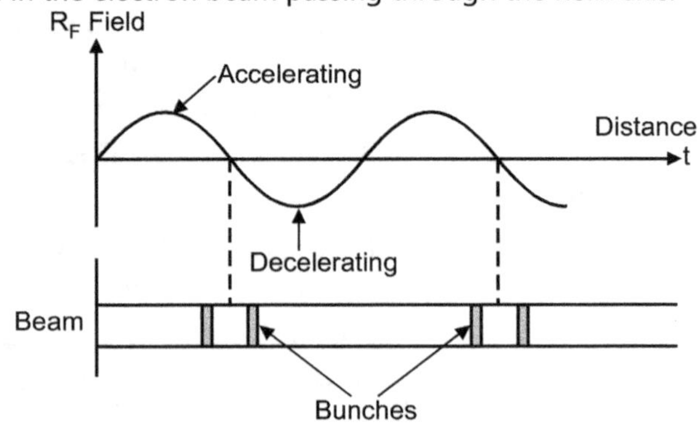

Fig. 1.13 : One wavelength

1.7.8 Advantages

1. It has ability to amplify a wide range of frequencies, i.e. a wide bandwidth.
2. Its power gain is high, about 40 to 70 dB.
3. Its output power is very robust, few watts to megawatts.
4. Its operating frequency is very high ranging from 360 MHz to 50 GHz.

1.7.9 Applications

1. It is used in Radar and other microwave receivers.
2. It is used in laboratory instruments.
3. It is used as a driver for more powerful tubes.
4. It is used for communication.
5. It is used for electronic counters.
6. It is used in Radar jamming.
7. It is used in air-borne and ship borne radars.
8. It is used in satellite and other space vehicles.

1.8 COMPARISON OF TWT AND KLYSTRON TUBES

Table 1.3

Klystron Tube	Travelling Wave Tube
1. It is a narrow band device.	1. It is a wide band device.
2. It uses cavity resonators for velocity modulation.	2. It uses non-resonant slow-wave structures for velocity modulation.
3. Interaction between electron beam and RF field occurs only across the cavity gap.	3. Continuous interaction between electron beam and RF field takes place.
4. The wave is not a propagating wave.	4. The wave is a propagating wave.
5. An average output power is 500 kW.	5. An average output power is upto 10 kW.
6. Power gain is about 30 dB.	6. Power gain is about 60 dB.
7. Efficiency is about 20 to 40%.	7. Efficiency is about 5 to 20%.
8. It is used as power output take in Radar transmitters and satellite ground stations, and also used as a power oscillator.	8. It is used as low noise amplifier for broadband applications in microwave receiver as a repeated amplifier in co-axial cable.

1.9 BACKWARD WAVE OSCILLATOR (BWO)

1.9.1 Introduction

- The Backward Wave Oscillator (BWC) were demonstrated in 1951, **M-type** and **Bernard Epsztein** and **O-type** by **Rudolf Kompfner**. The M-type BWC is a voltage controlled non-resonant extrapolation of **magnetron** interaction. Both types are tunable over a wide range of frequencies by varying the accelerating **voltage.**
- They can be swept through the band fast enough to be appearing to radiate over all the bands at once, which makes them suitable for effective **radar jamming**, quickly tuning into the radar frequency. Carcinotrons allowed airborne radar jammers to be highly effective. However, frequency-agile radars can help frequencies fast enough to force the jammer to use barrage jamming, during its output power over a wide band significantly impairing its efficiency.
- Concinotrons are used in research, civilian and military applications. For example, the **Czechoslovakia sensor** and **Ramona passive sensor** air defence detection systems employed carcinotrons at their receiver systems.

1.9.2 Definition

- A backward wave oscillator is abbreviated as BWC and also called carcinotron on backward wave tube.
- **A backward wave oscillator is a vacuum tube that is used to generate microwaves upto the tera Hertz (THz) range.**

$$1 \text{ THz} = 1 \times 10^{12} \text{ Hz.}$$

- Belonging to the travelling wave tube (TWT) family, it is an oscillator with a wide electronic tuning range.

1.9.3 Basic Concept

- The basic concept is dependent on a steady stream of electrons from an electron gun that travel down the center of the tube. Surrounding the electron beam is some sort of **radio frequency** source signal, in the case of the traditional Klystron. This is a resonant cavity fed with an external signal, whereas in more modern devices, there are a series of these cavities or a helical metal wire fed with the same signal.
- As the electrons travel down the tube, they interact with the RF signal. The electrons are attracted to areas with maximum positive bias and repelled from negative areas. This causes the electrons to bunch up as they are repelled or attracted along the length of tubes, and the process is known as **velocity modulation**. This process makes the electron beam take on the general structure as the original signal, the density of the electrons in the beam matches the relative amplitude of the RF signal in the induction system. The result is a signal in the electron beam that is an amplified version of the RF singal.
- As the electrons are moving, they induce a magnetic field in any nearby conductor. This allows the non-amplified signal to be extracted in systems like the magnetron klystron. This is accomplished with another resonant cavity. In the helical designs, this process occurs along the entire length of the tube reinforcing the original signal in the helical conductor.
- The **problem** with traditional designs is that they have relatively narrow bandwidths; designs based on resonators will work with signals within 10% or 20% of their designs, as this is physically built into the resonator design, while the helix designs have a much wider **bandwidth**, perhaps 100% on either side of the design peak.

1.9.4 Construction

- The construction of a blocking wave oscillator is shown in Fig. 1.14. It is made of tube, in which heater, cathode, helix, feedback conductor and focusing electrodes and anode are located.
- The waveguide cavity resonator is used for the trigger known as **butcher cavity resonator.** At the output end, there is another waveguide cavity resonator, which is known as **catcher cavity resonator** and used for the output of backward wave oscillator.

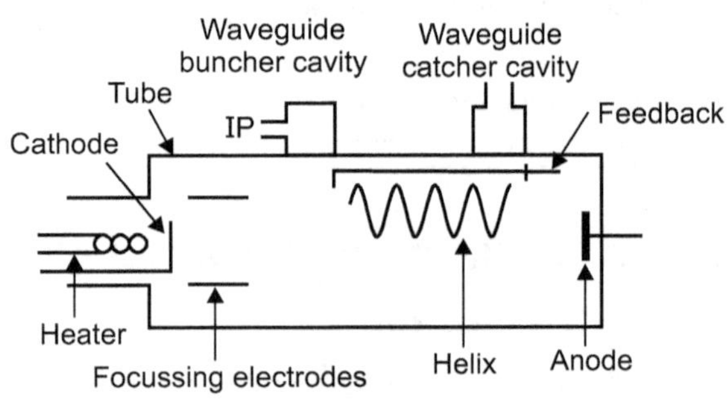

Fig. 1.14 : Backward wave oscillator

1.9.5 Principle of Working

- When we switch on the circuit, the heater will start to heat up. As a result, thermionic emission from cathode will start. The electrons will move from cathode and will be focussed by the focussing electrodes. It will be collected by this anode and it will travel back to the cathode in order to complete its path.
- When the input signal is applied to the **buncher cavity** the waveguide buncher cavity will excite. This will induce the voltages at the input end of the helix present in the form of the buncher cavity. As the induced voltage is A.C. signal in the helix, it will have positive and negative half cycles. If we consider the input negative half cycle at the helix, the electrons emitted from the cathode will be recalled back at the input of helix, which will result to form a bunch.
- Now for positive half cycles these electrons bunch will speed up and induce voltage in the next ring of the helix. This induced voltage will be stronger then the previous one. Finally, at the output end of the helix, the induced voltage will be so much strong that, it will excite the catcher cavity.
- The output will be obtained from the catcher cavity with the help of loop coupling. The feedback conductor is placed over the helix in order to overcome the damping effects and to maintain the oscillation.

1.10 CROSSED FIELD AMPLIFIER (CFA)

1.10.1 Introduction

- A cross-field amplifier is a specialized **vacuum tube**, first introduced in the mid-1950s and frequently used as a microwave amplifier in very high power **transmitters. Raytheon** engineer **William C. Brown's** work to adapt **magnetron** principles to create a new broadband amplifier is generally recognized the first CFA, which he called an **Amplitron**. Other names that are sometimes used by CFA manufacturers include **Platinotron** or **Sabiletron.**
- A CFA has lower **gain** and **bandwidth** than other microwave amplifier tube, such as **Klystron** or **travelling-wave tubes;** but it is more efficient and capable of much higher output power.
- Peak output powers of many mega watts (MW) and average power levels of tons of kilowatts (kW) can be achieved, with efficiency ratings in excess of 20% their current is in ground stations for TV rebroadcasting and deep space telecommunications networks.

1.10.2 Definition

- **The crossed field amplifier is a microwave power amplifier based on the magnetron and looking very much like it**.
- It is a cross between the travelling wave tube (TWT) and magnetron in its operation. It uses an essentially magnetron structure to provide interaction between crossed d.c. electric and magnetic **fields** and on RF **field.**
- **A crossed-field amplifier is a broadband microwave amplifier that can also be used as an oscillator (stabilitron).**
- It is so called **velocity modulated tube**. It is similar in operation to the magnetron and is capable of providing relatively large amount of power with high efficiency.

1.10.3 Principle of Working

- The constructional structure of a crossed field amplifier is shown in Fig. 1.15.
- The electric and magnetic fields in a crossed field amplifier are perpendicular to each other, i.e. cross fields. This is the same type of field interaction used in a magnetron; as a result, the two devices share many characteristics, such as high peak power and efficiency, and they have similar physical appearances. However, a magnetron is an oscillator and a crossed-field amplifier is an amplifier. RF circuit of a crossed-field amplifier or slow-wave structure is similar to that of coupled-cavity travelling wave tube (TWT).

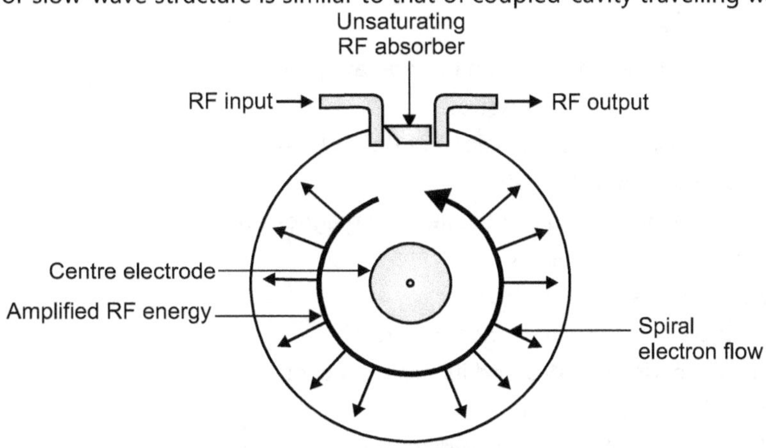

Fig. 1.15 : Crossed field amplifier

- The crossed-field amplifier has the useful property that when power is shut-off, the input simply passes to the output with very little loss. This avoids the need for bypass switching in the event of failure.
- Two crossed field amplifiers can be connected sequentially with only one powered, if it fails, power can be removed from the primary tube and applied to the secondary as a back up. This approach with built-in redundancy was used on the **S-band** downlink transmitter on the **Appollow Lunar Model** where high efficiency and reliability were needed.

- A large negative voltage is placed on the center electrode and large magnetic field is directed perpendicular to the page. This forms a thin spinning disk of electrons with a flow pattern like spinning water as it drains from a sink or toilet.
- A slow-wave structure is located above and below the spinning disk of electrons. Electrons flow much slower than the speed of light, and the slow wave structure reduces the velocity of the input RF enough to match the electron velocity.
- The RF input is introduced into the slow wave structure. The alternating microwave field causes the electrons to alternately speed up and slow down. These disturbances grow larger as electrons spiral around the device, and electrons slow down as the RF energy grows. This produces amplification. There is a small amount of RF feedback from output to input. This creates a slight random **phase jitter,** when the device is pushed.

1.10.4 Benefits or Advantages

1. Higher efficiencies upto 80% can be achieved.
2. Higher efficiencies are achieved without compromising on bandwidth.
3. Wide bandwidth can be achieved.
4. Bandwidth of 25% or less of centre frequency and as high as 50% of centre frequency can be achieved.
5. It has higher power handling capability, peak power of 4 MW at 2 GHz and 1 MW at 10 GHz can be achieved by using pulsed crossed-field amplifier.
6. It has slow wave structure similar to travelling wave tubes.

1.10.5 Drawbacks or Disadvantages

1. It has low gain about 10 to 15 dB.
2. It has high noise output.
3. It has low average power capability.

Practice Questions

1. What is microwave ? List its frequency range.
2. List microwave band designations with frequency range.
3. State important applications of microwave.
4. Classify microwave tubes.
5. State the principle and draw construction of two-cavity Klystron.
6. Describe the working of two-cavity Klystron.
7. State the working principle and applications of Reflex Klystron.
8. Explain working principle of a magnetron.

0\. Explain the effect of magnetic and electric fields in magnetron.

10. Describe with diagram working of TWT.
11. What is backward wave oscillator (BWO) ?
12. Describe construction of BWO with its neat diagram.
13. Explain principle of working of BWO and its basic concept.
14. What is crossed-field amplifier (CFA) ?
15. Describe construction of CFA with its neat diagram.
16. State benefits of crossed-field amplifier.
17. Explain principle of working of CFA.
18. State drawbacks of CFA.

✍ ✍ ✍

Chapter 2

MICROWAVE SOLID STATE DEVICES

Syllabus

Limitations of Conventional solid state devices at Microwaves, Transistors (Bipolar, FET), Diodes (Tunnel, Varactor, PIN), Transferred Electron Devices (Gunn diode), Avalanche transit time effect (IMPATT, TRAPATT, SBD), Microwave Amplification by Stimulated Emission of Radiation (MASER).

2.2 MICROWAVE SOLID STATE DEVICES

2.1.1 Introduction

- Semiconductor devices used for detection, generation, amplification, and control of electromagnetic radiation with wavelengths from 30 cm to 1 mm (frequency from 1 to 300 GHz). The number and variety of microwave semiconductor devices, used for wireless and satellite communication, and optoelectronics have increased as new techniques, materials, and concepts have been developed and applied.
- Passive microwave devices, such as PN and PIN junctions, Schottky barrier diodes, and varactors are primarily used for detecting, mixing, or controlling microwave signals. Stop recovery diodes, transistors, tunnel diodes and transferred electron devices (TEDs) are active microwave devices that generate power or amplify microwave signals.
- Typical high frequency semiconductor materials include silicon (Si), germanium (Ge), and compound semiconductors, such as germanium arsenide (GeAs), indium phosphite (InP), silicon germanium (SiGe), silicon carbide (SiC), and gallium nitride (GaN). In general, the compound semiconductors work best for high-frequency applications due to their higher electron mobilities.

2.1.2 Classification

(A) Depending upon their electrical behaviour :

1. Non-linear resistance type : Varistors
2. Non-linear reactance type : Varactors
3. Negative resistance type : Tunnel diode, IMPATT diode, Gunn diode
4. Controllable impedance type : PIN diode

(B) Depending upon their construction :

1. Point contact diodes
2. Schottky barrier diodes
3. MCS devices
4. Metal insulation devices.

2.1.3 Limitations

1. Stray capacitances
2. Lead inductances
3. Transit time
4. Reduced efficiency

2.2 MICROWAVE BJT

2.2.1 Introduction

- Silicon Bipolar Junction Transistors (BJT) were first on the microwave scene, followed by GaAs field-effect transistors (FETs). Indeed, FETs now have noticeably lower noise figures, and in the C-band and above they yield noticeably high powers.

2.2.2 Definitions

- Microwave bipolar junction transistor (BJT) is one of the axial planar silicon n-p-n type transistor operating at microwave frequencies.

2.2.3 Requirements

- The various factors that contribute to a maximum high-frequency performance of microwave transistors are complex. They include the already mentioned requirements for high voltages and currents, and two other conditions as under :
  1. Small electrode area to reduce inter-electrode capacitance.
  2. Very narrow active regions to reduce transit time.
  3. Smaller emitter junction.
  4. Very thin base.

2.2.4 Construction

- Silicon planar transistors offer the best bipolar microwave transistor's performance. Fabrication difficulties, together with the excellent performance of GaAs FETs, have prevented the manufacture of GaAs bipolars. Epitaxial diffused structurers are used, giving a combination of small emitter area and large emitter edge.
- The first property gives a short transit time through the emitter, and the second a large current capacity. The inter-digitated transistor is the most common bipolar in production. The transistor shown has a base and emitter layout that is similar to two bands with interlocking fingers, hence its name.
- The chip illustrated have overall dimensions (less contacts) of about 70 × 70 μm, the emitter contact is on the left, the base on the right and the collector underneath. The thickness of each emitter and base **"finger"** in the transistor shown is 0.5 μm. This yields values of f_{max} (maximum frequency) in excess of 20 GHz, 0.25 μm geometries have been proposed.

2.2.5 Applications

1. Power amplifiers
2. Oscillators
3. Output stages in microwave links
4. Driver amplifiers.

2.3 MICROWAVE FET

2.3.1 Introduction

- **Microwave field-effect transistor (FET) is generally a Schottky barrier gate type gallium arsenide (GaAs) field effect transistor, operating at microwave frequencies.**
- Gallium arsenide is preferred to silicon because its electron mobility is greater than that of silicon.

2.3.2 Construction

- The most common microwave FET uses a **Schottky-barrier gate**. Its construction is shown in Fig. 2.1 which demonstrates why this device is known as **metal semiconductor field effect transistor** (**MESFET**). The cross-section shows it to be a **mesa** type construction. The top metallic layer has been ratched away as it has a portion of n-type GaAs semiconductor underneath. The metallic Schottky-barrier gate strip is deposited in the resulting groove.
- It has a typical length of 1 μm (the normal range is 0.5 to 3 μm). The width of the gate is not shown in the cross-section, 300 to 2400 μm is a typical range. Dual-gate GaAs FETs are also available, in which second gate may be used for application of AGC in receiver RF amplifiers. It should be mentioned that values of maximum frequencies f_{max} in excess of 100 GHz are currently achievable.

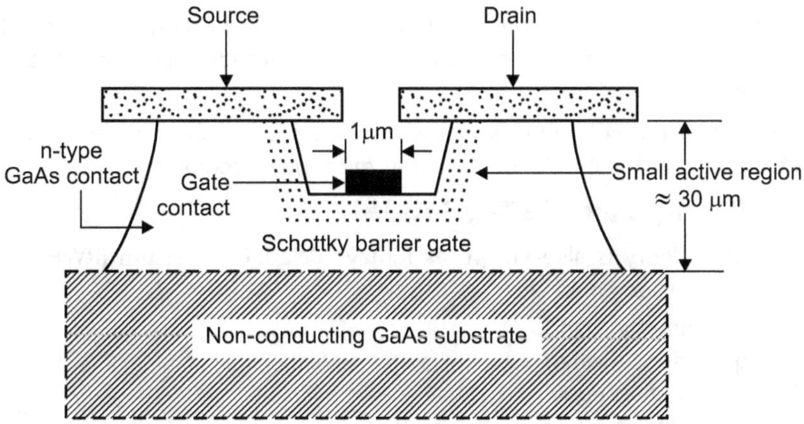

Fig. 2.1 : Microwave semiconductor field effect transistor (MESFET)

2.3.3 Advantages

1. High electron mobility.
2. Low capacitance levels.
3. High input impedance.
4. Negative temperature coefficients.
5. Lack of oxide traps.
6. High level of geometry control.

2.3.4 Applications

1. Power amplifiers.
2. Oscillators.
3. Microwave links.
4. Driver amplifiers.
5. High power transmitters.
6. Base array radars.
7. Broadband generators.

2.3.5 Schottky Contacts

- **Schottky contact is a diode junction formed between certain metals and semiconductors.** This contact is named by Mr. Schottky. Metals that form Schottky contacts to N-type GaAs include aluminium, gold, silver, titanium and platinum. Often a layered structure of metals is used in FETs, such as titanium, platinum, gold.
- Gold reacts with GaAs, so it is a band Schottky metal. Platinum keeps the gold away from GaAs. It acts as a barrier metal in this case. Titanium make stick to GaAs.

2.4 GUNN DIODE

2.4.1 Introduction

- High microwave power cannot be obtained from junction diodes, because at high frequencies the junction width goes on thinning. In 1963, the scientist *J. B. Gunn* discovered a new phenomenon called *transferred electron effect* while studying the properties of thin specimens of Gallium Arsenide and Indium Phosphide under high electric stress.
- The new effect is now known as *Gunn effect*, after the name of the scientist, who discovered it. A very popular microwave diode is a *Gunn diode*. It is also known as *Transferred Electron Device* (TED).

2.4.2 Gunn Effect

- It should be noted that the Gunn Effect is a bulk property of semiconductor and is quite independent of their junction or contact properties. Also the effect is independent of total voltage or current and it is not affected by the magnetic field.
- Bulk effect is only found in n-type materials, so that it must be associated with the electrons rather than holes. If a small d.c. voltage is applied across a thin slice of Gallium Arsenide, a negative resistance is encountered under certain conditions.

- Thus, in this phenomenon, the electrons are transferred from the conduction band to a higher energy band in which they are much less mobile and thus the current reduces as a result of increase in voltage to give the negative resistance region.
- As the slice is very thin, the voltage gradient across the slice of GaAs is very high. Hence, the electron velocity is also high, so that the oscillations occur at microwave frequencies.

2.4.3 Gunn Effect Using Two Valley Model

- The two valley model theory is also known as Ridlay - Watkins - Hilsum (RWH) theory. RWH theory is based on *population inversion principle*.
- For an n-type GaAs energy band model under equilibrium, when electric field E is greater than equal to threshold value (3.3 kV/cm) and the electrons are transferred from lower valley to upper valley of conduction band in which they are much less mobile and thus, the current reduces, and the device offers negative resistance. This is also known as *transferred electron effect*.

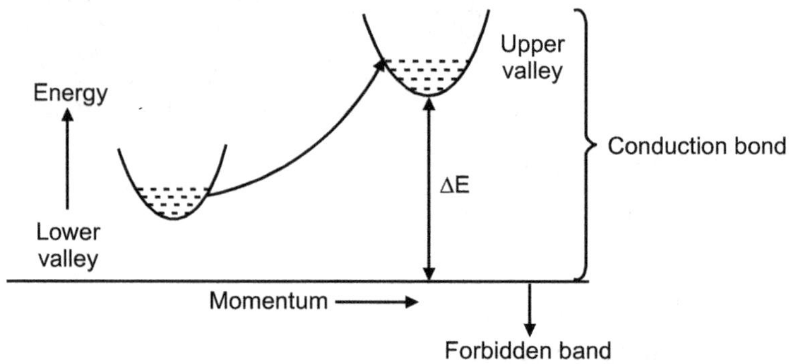

Fig. 2.2 : Two valley model of GaAs

- Fig. 2.2 shows a two valley model of n-type GaAs.
- The conductivity (σ) of n-type GaAs is given by,

$$\sigma = e(\mu_l R_l + m_a R_u) \quad \ldots (2.1)$$

where,

e = Charge on electron

μ_l = Mobility of electron in lower valley

μ_u = Mobility of electron in upper valley

R_l = Density of electron in lower valley

R_u = Density of electron in upper valley

2.4.4 Basic Principle

- A Gunn diode is so biased that the time it takes for electrons to flow across the Gallium Arsenide material, such that the current will be at 180°C out of phase with applied voltage.

2.4.5 Salient Features

1. It is available to oscillate at frequencies upto about 50 GHz.
2. In the lower microwave, power output upto watts are possible.
3. The thickness of semiconductor, determines the frequency of oscillation.
4. If the cavity is made variable, the oscillator frequency can be adjusted over a narrower range.

2.4.6 Construction

- The Gunn diode is not a diode in the usual sense, as it does not have a semiconductor junction.
- The construction of a Gunn diode is shown in Fig. 2.3.
- It consists of a thin piece, i.e. slice of an n-type Gallium Arsenide (GaAs) or Indium Phosphide (InP) semiconductor.
- It has a buffer layer between the active layer and the substrate.
- It is grown epitaxially out of GaAs or InP doped with Silicon, Tellurium or Selenium.

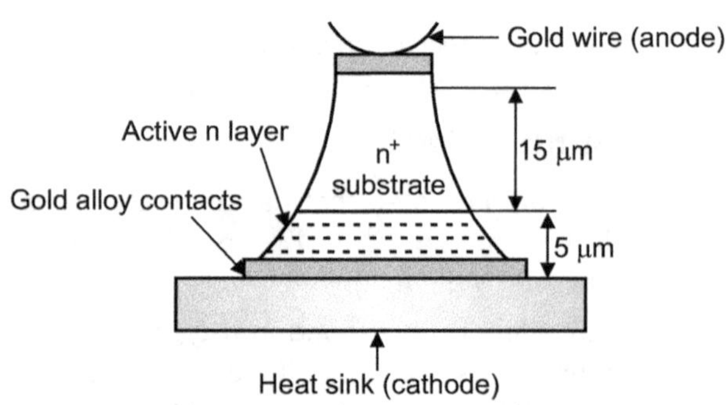

Fig. 2.3 : Epitaxial GaAs Gunn diode

- The substrate used here as an ohmic contact is highly doped for good conductivity, while thin active layer is less heavily doped. The gold alloy contacts are electrodeposited and used for good ohmic contact and heat transfer for subsequent dissipation. The active layer varies in thickness depending on the frequency.

2.4.7 Principle of Operation

- Consider the structure of a Gunn diode as shown in Fig. 2.3. If a relatively small d.c. voltage is applied across a thin slice of Gallium Arsenide (GaAs). On application of voltage across an n-type GaAs slice, electrons flow as a current towards the positive end of the slice.
- The greater the potential across the slice, the higher is the velocity with which the electrons move towards the positive end, and therefore, greater current. Now the device behaves as a normal positive resistance. The high voltage gradient imparts much energy to the electron, that instead of travelling faster and, therefore, constituting a larger current, they actually slow down.
- This happens due to the very high energy imparted to the electrons so that they get transferred into the empty energy band and so get slowed down. This gives rise to the transferred electron effect. The electrons are thus transferred from the conduction band to a higher energy band in which they are much less mobile and so the current gets reduced as a result of increased voltage.
- As the voltage rises past the threshold level, current falls and thus negative resistance is exhibited by the GaAs slice. As the voltage rises further, the gradient becomes sufficient to remove electrons from the higher energy lower mobility band, so that current again starts rising with voltage.

2.4.8 Characteristics

- When a d.c. bias voltage applied to Gunn diode is increased the current through it also increases gradually.
- When the d.c. bias increases sufficiently to produce electric field of 2.4 kV/cm, the current decreases with increase in d.c. bias. This results in negative resistance effect. The V-I characteristics of a Gunn diode is as shown in Fig. 2.4.

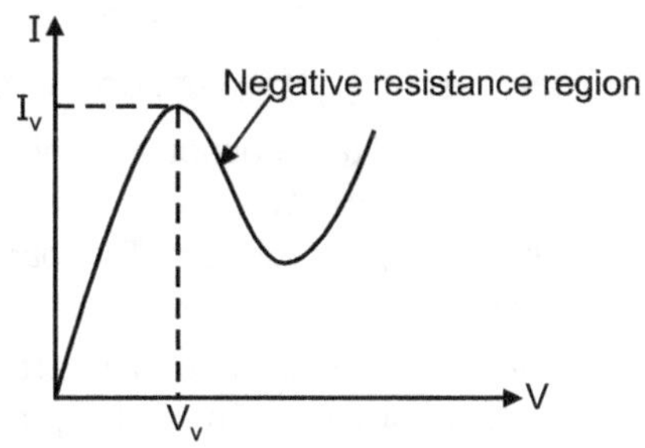

Fig. 2.4 : Characteristics of a Gunn diode

2.4.9 Applications

1. It is used in variety of low power transmitter applications.
2. Its oscillations are used in police Radar, CW Doppler Radar, Burglar alarms and air-crafts rate of climb indicators in broadband.
3. It is used in parametric amplifiers as a pump source.
4. It is used in broadband microwave amplifiers.
5. It is used as a low and medium power oscillators in microwave receivers.

2.4.10 Gunn Diode as an Oscillator

- Fig. 2.5 shows two oscillator circuit arrangements using a Gunn diode. Fig. 2.5 (a) shows a typical co-axial Gunn oscillator.
- If some electrical tuning is required, a varactor diode may be placed in the cavity at the opposite end to the Gunn diode.
- When the frequency of the resonator is changed, the Gunn diode itself responds by generating its domain at a distance from the anode such that the transit time of the domain corresponds to a cycle of oscillation. Thus, as the resonating frequency of the tuned circuit is raised, that domain formation point moves towards the anode. The oscillations stop when the domain formation point is more than half way across the slice.

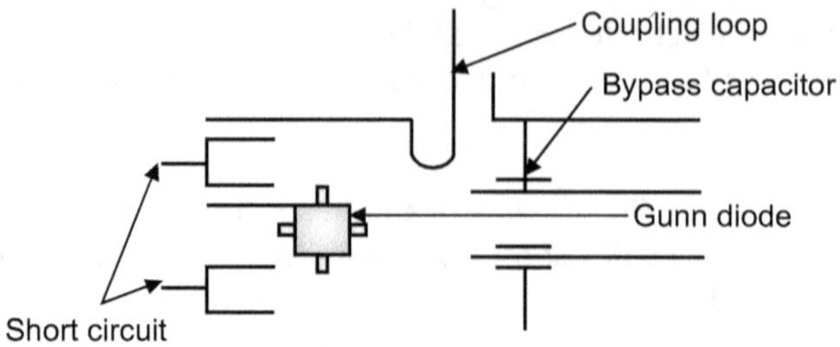

(a) Coaxial type

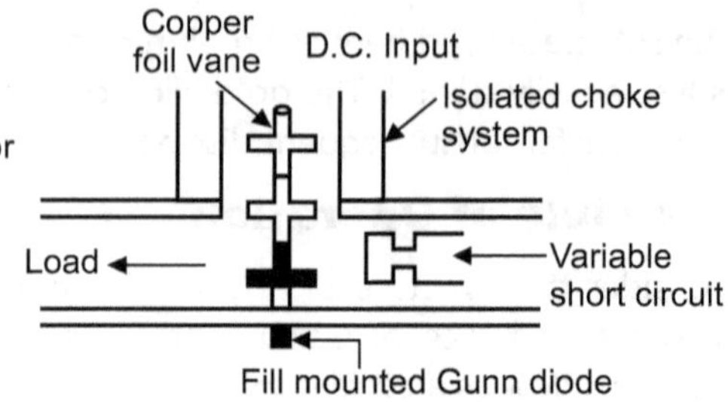

(b) Waveguide type

Fig. 2.5 : Gunn diode oscillator

2.5 ESAKI OR TUNNEL DIODE

2.5.1 Introduction

- If the concentration of impurity atoms is greatly increased say, to 1 part in 10^3 in a P-N junction, then the width of the depletion region reduces to about 100 A° and hence the diode characteristics are completely changed. This gives rise to a new type of P-N junction diode.
- A Japanese scientist *Dr. Leo Esaki* observed/explained a peculiar characteristic in forward direction in case of a P-N junction diode in 1957. In 1958, he announced the new junction diode and gave appropriate theoretical explanation for its V-I characteristics.
- This diode utilizes the phenomenon called *tunneling* and hence, this diode is referred to as tunnel diode. After the name of an inventor, the tunnel diode is also referred to as a *Esaki diode*.

2.5.2 Tunneling Phenomenon

- The width of the junction barrier varies inversely as the square root of the impurity concentration. Hence, with large impurity concentration of the order of 1 part in 10^3, the junction barrier width reduces to less than 100 A°, i.e. 10^{-6} cm. This barrier width is very small being only about one fifth of the wavelength of the visible light.
- As per the classical theory, a particle must have an energy atleast equal to the height of potential energy barrier in order to cross the junction barrier. However, if the junction barrier is extremely thin about 100 A°, such as encountered in a tunnel diode, there is a non-zero probability than the electron may penetrate through the barrier at the speed of light.
- As a result of this, a large forward current is produced even if the applied forward voltage is much less than the knee voltage, i.e. 0.3 V. This quantum mechanical behaviour is referred to as tunneling and the high impurity density P-N junction device is known as a *tunnel diode*. The phenomenon of penetrating the charge carriers directly through the potential barrier instead of climbing over it, is called as *tunneling*.

2.5.3 Construction

- Fig. 2.6 shows the constructional features of a typical tunnel diode.
- A very small tin dot of about 50 μm in diameter is soldered or alloyed to a heavily doped pallet of n-type Ge, GaSb or GaAs.
- The pallete is then soldered to a Kovor base, which is used for heat dissipation and as anode contact.
- The cathode contact is obtained by connecting the tin dot to a Kovar cap through a mesh screen, which is used for reducing the inductance.
- The outer body of the diode is made of ceramic having hermetically sealing lid on the top.

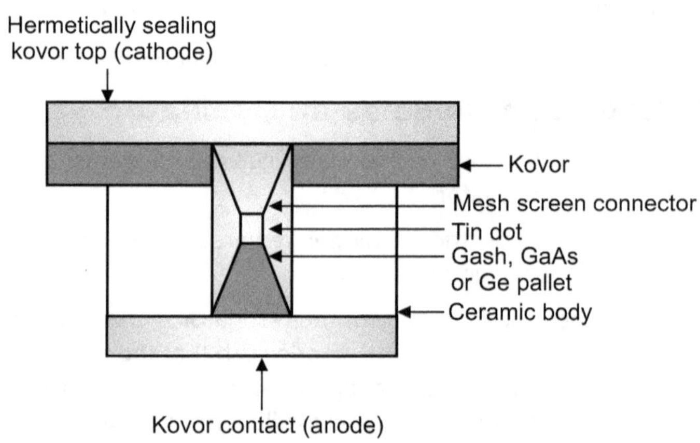

Fig. 2.6 : Tunnel diode

2.5.4 Principle of Operation

- The energy possessed by the electrons on the n-side of the junction, under no bias conditions, is insufficient to permit them to jump over the junction barrier and reach the p-side. However, the quantum mechanics proves that there exists very small but finite probability that an electron with insufficient energy to climb the barrier can, nevertheless find itself on the other side of it.
- If this junction barrier is thin enough, without any loss of energy on the part of the electron. An electron on one side of the junction barrier will have a certain probability of leaking through the junction barrier, if the junction barrier is very thin of the order of 100 A°. If both the p and n materials of a junction are made heavily doped, the depletion region becomes very narrow.
- It is possible to make the thickness of the junction barrier as narrow as of the order of 100 A°. Under unbiased conditions, net tunneling of the thin barrier is zero. As forward bias is applied across the junction, the electrons climb the thin barrier and tunneling takes place. The tunneling current increases, reach its maximum value at a forward bias of the order of 0.1 V.
- As the forward bias is further increased beyond certain level, almost all electrons climb the barrier. This stops the tunneling action and the normal forward current will reduce for increased forward voltage. This exhibits negative resistance. When the forward bias is increased still further normal forward current flows and now the tunnel diode behaves as an ordinary rectifier diode.

2.5.5 V-I Characteristics

- The voltage current (V-I) characteristics of a typical tunnel diode is shown in Fig. 2.7. The forward current rises sharply as the forward voltage is applied.
- At point A, the peak voltage occurs and the forward current is maximum. As the forward bias is increased still further, the forward current drops very rapidly and continues to drop until point B is reached. This is the valley voltage.
- At the point B, the forward current starts to increase once again and does so very rapidly as forward bias is further increased. The tunnel diode exhibits dynamic negative resistance between the points A and B.

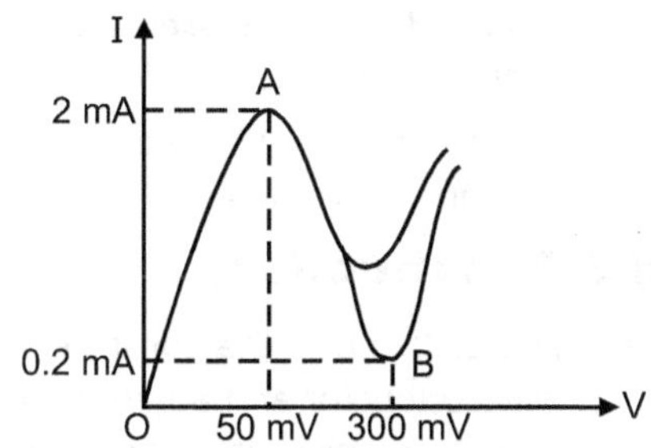

Fig. 2.7 : V-I characteristics of a tunnel diode

2.5.6 Equivalent Circuit

- Fig. 2.8 shows a small signal equivalent circuit of a typical tunnel diode. It consists of a series resistance R_s, an inductance L_s, a junction capacitance C_j and a negative resistance R_n.
- The series resistance is the resistance due to loads, contacts and semiconductor material.
- The junction capacitor C_j is due to diffusion capacitance and the applied voltage. It is a function of the applied voltage and it is measured usually at the valley point.

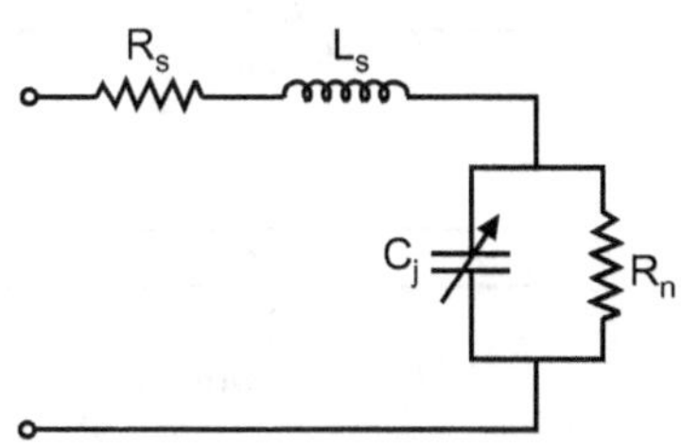

Fig. 2.8 : Equivalent circuit of a tunnel diode

2.5.7 Applications

1. It is used in relaxation oscillator.
2. It is used in high speed switching.
3. It is used in logic operations as flip-flop and gates.
4. It is used as a microwave amplifier.
5. It can be used as a mixer and frequency converter.
6. It can be used as a microwave limitation.

2.5.8 PIN Diode as a Modulator

- The PIN diode is kept at low reverse bias and in series with the low frequency modulating signal. The modulating signal amplitude is kept smaller than the RF carrier signal.
- The modulating signal changes the RF resistances of PIN diode, so that varying amount of mismatch results.

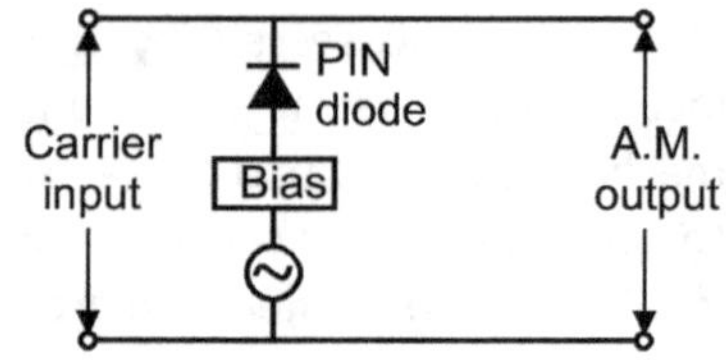

Fig. 2.10 : PIN diode as a modulator

- Hence, the amount of carrier power is reflected back and so the amount of carrier is passed beyond the PIN diode circuit towards the output variacs as the input value of modulating signal.

2.6 PIN DIODE

2.6.1 Introduction

- The name of the diode as PIN is derived from its construction. PIN diode was first proposed in R.N. Hall in 1952, while its use as a microwave switch was first recognized in 1958 by Uhlir.

2.6.2 Definition

- **A PIN diode is a useful control element at microwave frequencies.** It differs considerably from an ordinary P-N junction diode.
- It is also a microwave passive diode. It is only passive only to the extent it is not used in power generation or amplification apart from that, it is very active indeed in mixers, detectors and power control.

2.6.3 Construction

- Fig. 2.10 shows the construction of a PIN diode. It consists of a narrow layer of a heavily doped p-type semiconductor and a narrow layer of a heavily doped n-type semiconductor separated by a thinner layer of high resistivity intrinsic material as shown in Fig. 2.10 (a).

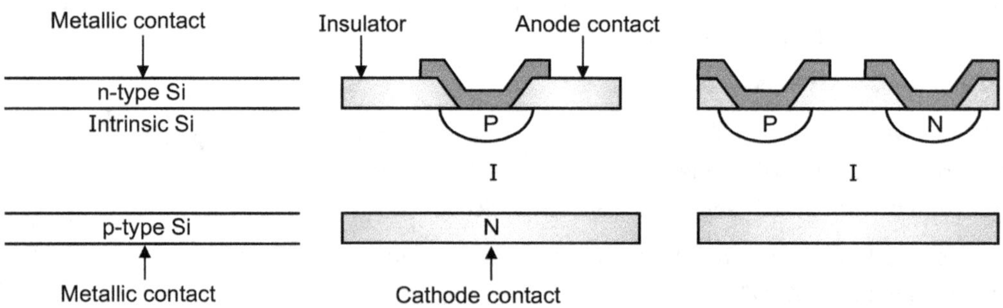

(a) Schematic diagram **(b) Planar construction** **(c) Planar diode with in-line orientation**

Fig. 2.10 : PIN diode

- The thickness of the high resistivity layer is usually in the range of 10 μm to 200 μm. The intrinsic layer has a resistivity of about 1 kΩ-cm and is either p-type or n-type. This lightly doped layer is called *π-type* or *v-type* depending upon whether the conductivity is *p-type* or *n-type* respectively.
- The planar construction is used because it gives a lower series resistance while conducting. The in-line construction of a diode reduces the diode shunt capacitance. Silicon is the main material used in construction, but sometimes Gallium Arsenide is also used.

2.6.4 Principle of Operation

- At zero bias, the diffusion of holes and electrons across the junction causes space charge region of thickness inversely proportional to impurity concentration. As 'i' layer has no depletion region, i.e. p-layer has a fixed negative charge and n-layer has a fixed positive charge under zener bias.

- As the reverse bias is applied, the space regions in the p- and n-layer will become thicker, as the carriers in p- and n-layers get attracted towards the battery. The reverse resistance will be very high and remains almost constant.
- When forward bias is applied, the carriers will be injected into the 'i' layer from p-side to n-side.
- This results in carrier concentration in the 'i' layer and it raises the carrier concentration above equilibrium level. Thus, the resistance decreases as forward bias is increased. So, most of the power is reflected and only a very small part is transmitted. Thus, the PIN diode acts as an electronic switch.
- The PIN diode acts as an ordinary rectifier diode upto about 100 MHz. However, above this frequency, it ceases to behave as an ordinary rectifier diode because of the carrier storage in and the transit time across the intrinsic region. At microwave frequencies, the PIN diode acts as a variable resistor.
- When a forward bias is varied on the PIN diode, its microwave resistance changes from a typical value of 0.5 Ω to 10 Ω, but under negative bias conditions, the microwave resistance changes from a typical value of 5 kΩ to 10 kΩ. Thus, if a PIN diode is mounted across a 50 Ω coaxial line, it will not significantly load the line, when it is reverse biased, so that the power flow is unaffected.

2.6.5 Equivalent Circuit

- When the PIN diode is mounted across a 50 Ω co-axial line, it will not lead the line under reverse bias. When it is forward biased, it reflects most of the power due to mismatch and leading.
- To improve the power handling capacity, several diodes can be used in parallel and switching of more than 150 kW peak power can be achieved.
- The shunt depletion capacitor C_j limits the upper frequency operation. The equivalent circuit of a PIN diode is shown in Fig. 2.11.

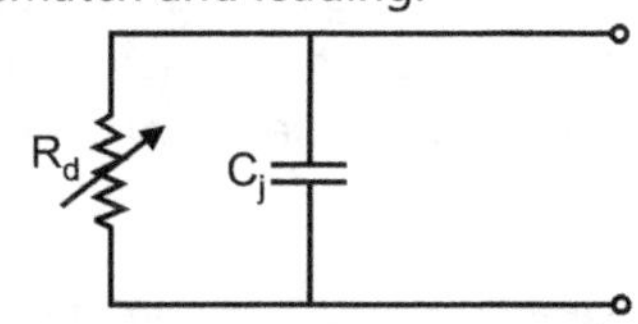

Fig. 2.11 : Equivalent circuit of a PIN diode

2.6.6 Characteristics

- The characteristics of a PIN diode is as shown in Fig. 2.12.
- It shows the variation of resistance with increase in forward bias.
- At zero bias, the resistance is maximum about 10 kΩ and with increase in forward bias, the resistance reaches to minimum of about 5 Ω.

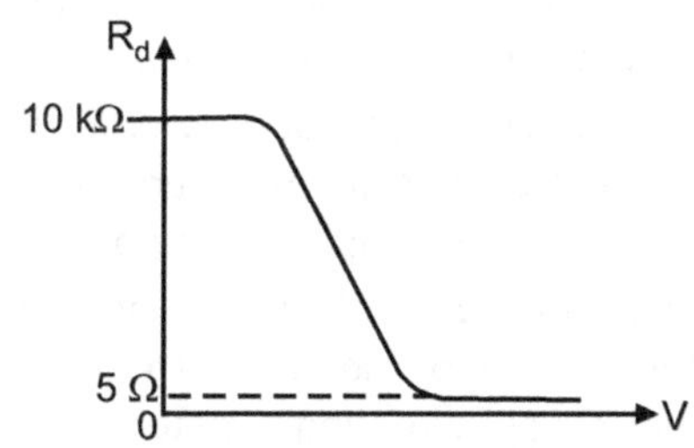

Fig. 2.12 : Resistance variation with bias

2.6.7 Applications

1. It is used as a square wave modulator and a pulse modulator.
2. It is used as a switch, a phase shifter and a limiter.
3. It is used as a voltage controlled attenuator.
4. It is used in mixers, detectors and power controllers.

2.7 VARACTOR DIODE

2.7.1 Definition

- **A semiconductor diode whose internal capacitance varies with the variation of the reverse voltage, such type of semiconductor diode is known as varactor diode.**
- The varactor diode is also known as **varicap**. It is used for storing the charge. It always works in reverse bias, and it is a voltage dependent semiconductor device.
- **The varactor diode is a voltage dependent semiconductor diode whose internal capacitance varies with the variation of the reverse voltage across the junction.**

2.7.2 Symbol

- The symbol of a varactor diode is shown in Fig. 2.13. It is similar to that of the PN junction diode. It has two terminals namely anode A and cathode K. The one end of a symbol consists of a diode and their end has two parallel lines that represent the conductive plates of the capacitor. The gap between the plates shows their dielectric.

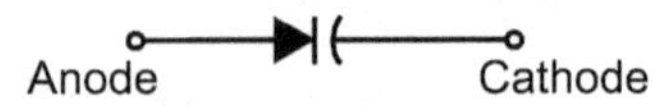

Fig. 2.13 : Varactor diode symbol

2.7.3 Construction

- The varactor diode consists of p-type and n-type semiconductors. In the p-type semiconductor, free electrons are the majority carriers and holes are the minority carriers. So, the free electrons carry most of the electric current in the n-type semiconductor.
- In the p-type semiconductor, holes are the majority carriers and free electrons are the minority carriers. So, the holes carry most of the electric current in p-type semiconductor.
- When a p-type semiconductor is in contact with the n-type semiconductor, a p-n junction is formed between them. The p-n junction separates the p-type and n-type semiconductors. At the p-n junction, a depletion region is created. A depletion region is a region, where mobile charge carriers, i.e. free electrons and holes, are absent.
- The depletion region is made up of positive and negative charged atoms, i.e. ions. These positive and negative ions does not move from one place to another place. The depletion region blocks free electrons from n-side and holes from p-side. This depletion region blocks electric current across the p-n junction.

2.7.4 Working

- The varactor diode is made of **n-type** and **p-type semiconductor** materials, the electrons are the majority charge carriers and in the p-type material, the holes are the majority carriers.
- When the p-type and n-type semiconductor materials are joined together, the p-n junction is formed, and the depletion region is created at the p-n junction. The positive and negative ions make the depletion region. This region blocks the current to enter from the p-n region.
- The varactor diode operates only in reverse bias. Because of reverse bias, the current does not flow. If the diode is connected in forward bias, the current starts flowing through the diode and their depletion region decreases. The depletion region does not allow the ions to move from one place to another.
- The capacitance of varactor diode increases with the increase of n and p-type regions and decreases with the increase of the depletion region. The increase in capacitance means the more charges are stored in the diode. For increasing the storage capacity of charge, the depletion region, which acts as a dielectric of the capacitor of the diode should be kept small.

2.7.5 Characteristics

- The characteristic curve of the varactor diode is shown in Fig. 2.14. The graph shows that when the reverse bias voltage increases, the depletion region increases and the capacitance of the diode reduces.

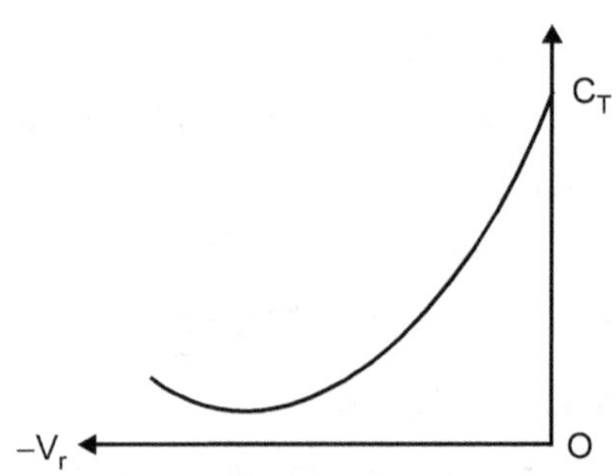

Fig. 2.14

2.7.6 Advantages

1. It produces comparatively less noise.
2. It is less costly.
3. It is more reliable.
4. It is small in size and less in weight.

2.7.7 Applications

1. It is used for storing of charge.
2. It is used in tuning circuit.
3. It is used as a voltage controlled capacitor.
4. It is used in voltage controlled capacitors, parametric amplifiers and frequency multipliers.
5. It is used in RF switching, RF filters and harmonic multiplication circuits.
6. It is used in AFC, RF phase shifter, and frequency multiplier circuits.
7. It is used in small remote control, modulation and demodulation circuits.

2.8 IMPATT DIODE

2.8.1 Introduction

- The long form of the IMPATT diode is Impact Avalanche and Transit Time Diode.
- Tunnel diodes exhibit the dynamic d.c. negative resistance, but it does not exhibit it for alternating current.
- For this purpose, an IMPATT diode was first developed in 1965. An IMPATT diode exhibits the negative resistance for alternating currents also.

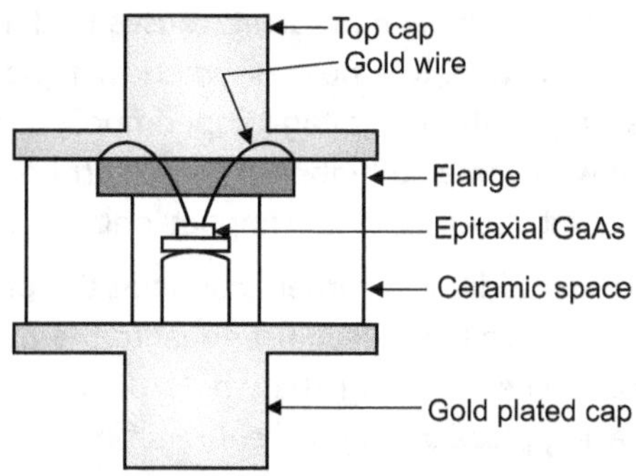

Fig. 2.15 : Construction of IMPATT diode

2.8.2 Definition

- **The microwave diode, exhibiting the negative resistance for alternating current due to Avalanche effect is known as an IMPATT diode.**

2.8.3 Avalanche Effect

- The negative resistance in the microwave diode can be caused by another method. For this purpose, delay is created between the voltage and the avalanche current and along with it transit time through the semiconductor material is provided to create negative resistance.
- A combination of delay involved in generating the avalanche current multiplication, together with delay due to transit time through a drift space provides the necessary 180° phase difference between the applied voltage and the resulting current in an IMPATT diode. Thus, *negative resistance* is exhibited by the IMPATT diode.

2.8.4 Construction

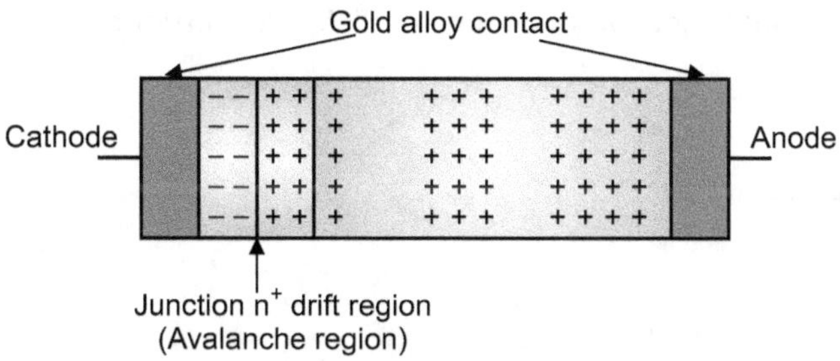

(a) Cross-section

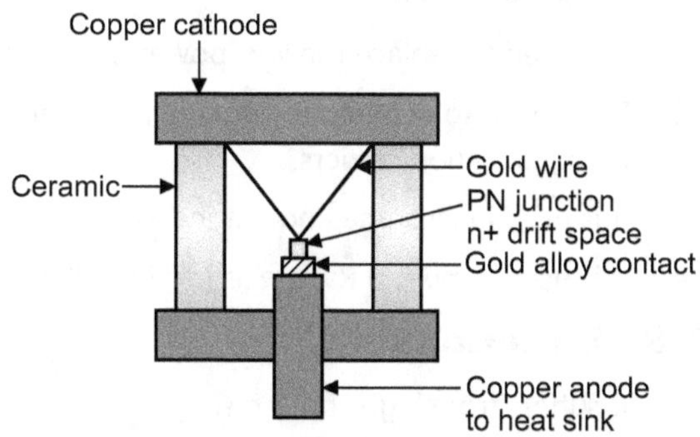

(b) Construction

Fig. 2.16 : IMPATT diode

- Fig. 2.16 shows the cross-sectional area of the active region of an IMPATT diode. As an IMPATT diode is a junction diode, there is a junction formed between the p^+ and n type semiconductor layers. Its cross-sectional area is very small, to ensure very small capacitance.
- It uses either a chip made of GaAs or silicon. Particularly the contacts have extremely low ohmic and thermal resistances.
- It is generally embedded in the walls of a cavity, which forms its natural heat sink.

2.8.5 Principle of Operation

- Consider the construction of an IMPATT diode as shown in Fig. 2.16 (b). An extremely high voltage gradient of the order of 400 kV/cm is applied to the IMPATT diode, which results in a very high current and generates the power of the order of 10 MW/cm^3. Under such high voltage gradient, it does not quickly breakdown, but it withstands such conditions repeatedly.
- This high voltage gradient, causes back biasing of the diode, resulting in a flow of minority carriers across the junction. It is now assumed that the oscillations already exist, we may consider the effect of positive swing of the RF voltage superimposed on top of the high d.c. voltage. Electron and hole velocities have now become so high that these carriers from additional holes and electrons by knocking them out of the crystal structure by the impact ionization.
- These additional carriers continue the process at the junction till avalanche takes place. If the original d.c. biasing voltage is set just equal to the threshold required for this avalanche then this threshold voltage will be exceeded using the whole of the positive RF cycle and the avalanche current multiplication will be taking place during this entire time.
- Since, it is a multiplication process, the avalanche is not instantaneous. The process takes a time such that the current pulse is maximum at the junction, at the instant upon the RF voltage across this diode is zero and going negative.
- A 90° phase difference between the voltage and the current is obtained. Because of the reverse bias, the current pulse does not stay at the junction, but it flows to the cathode at a drift velocity dependent on the presence of the high d.c. field.
- When the current pulse actually arrives at the cathode terminal, the RF voltage is at its negative peak. Accordingly, the voltage and current obtain a 180° phase difference between one another, thus giving the dynamic RF negative resistance.

2.8.6 Salient Features

1. It is very sensitive.
2. It is essentially a narrow band device, when used in amplifier.

2.8.7 Applications

1. It is used to replace the low-power magnetrons used in CW Radar and Electronic Counter Measures (ECM).
2. It is used to replace a large number of TWT transmitters working upto 10 W used in long distance communication carriers.
3. It is used in airborne and marine Radars.
4. It is used in pulsed Radars as local oscillator and in altimeter.

2.8.8 Limitations

1. It suffers from high shot noise.
2. It has poor efficiency.
3. It does not have the capability of being tuned through a wide frequency range.

2.9 TRAPATT DIODE

2.9.1 Introduction

- The TRAPATT stands for TRApped Plasma Avalanche Triggered Transit. It belongs to the same family as the IMPATT diode, but it provides number of advantages in some applications.
- The IMPATT diode is normally used as a microwave oscillator. It has the advantage of greater level of efficiency as compared to IMPATT microwave diode. Typically, the D.C. to RF signal conversion efficiency may be in the range of 20% to 60%, which is particularly high.

2.9.2 Definition

- **The TRAPATT diode is a microwave diode belonging to a family of an IMPATT diode, which differs higher level of efficiency.**
- **The TRAPATT diode is a high efficiency generator competent of operating from numerous hundred MHz to several GHz of frequency.**

2.9.3 Structure/Construction

- The structure of a TRAPATT diode is shown in Fig. 2.17, which is similar to a bipolar junction transistor (BJT). High peak power diodes are typically silicon n^+-p-n^+ or p^+-n-n^+ structures with the n-type depletion region with varying from 2.5 to 12.5 μm.
- The doping of depletion region is generally such that the diodes are well punched through at breakdown. The device p^+ region is kept as thin as possible at 2.5 to 6 μm. The TRAPATT diodes diameter ranges from a small as 50 μm for microwave operation to 750 μm at lower frequency for high peak device.

2.9.4 Principle of Operation

- A high field avalanche zone propagate through the diode and fills the depletion layer with a dense plasma of electron and holes that become trapped in low field region behind the zone. A typical voltage for the TRAPATT mode of an avalanche p^+-n^--n^+ diode operating with as assumed square wave current drive is shown in Fig. 2.17.
- At point A, the electric field is uniform throughout the sample and its magnitude is large, but less than the value required for avalanche breakdown.

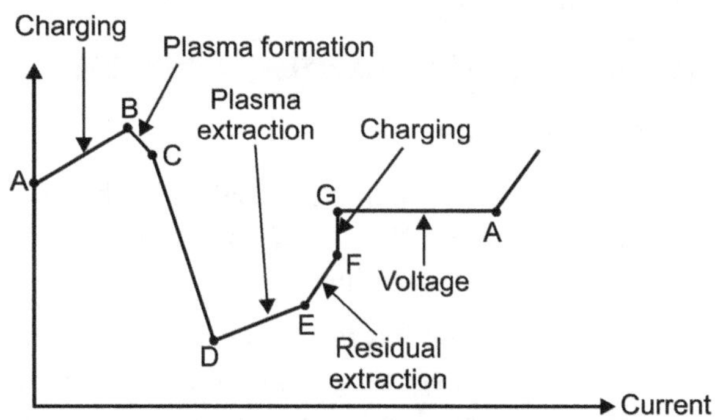

Fig. 2.17 : Characteristics of TRAPATT

- At the instant of time at point A, the diode current is turned on, since the charge carriers present are those caused by thermal generation. The diode initially charges up like a linear capacitor, driving the magnitude of the electric field above the breakdown voltage.
- When sufficient number of carriers are generated, the particle current exceeds the external current and the electric field is depressed throughout the depletion region, causing the voltage to decrease. This portion of cycle is known by the curve point B to point C. During the time interval, the electric field is sufficiently large for the avalanche to continue and a dense plasma of electrons and holes is created. The voltage decreases to the point D.
- The holes produced in the avalanche rapidly reach the p^+ contact taking no part in the process, but the electrons are released into the n-region, where they do not combine with either donor or holes.
- The electron drifts at their maximum velocity across the region and current continues to flow in the external circuit, which they are in transit. When this current pulse actually arrive at the cathode terminal, the a.c. voltage is at its negative peak and the second delay of 90° has taken place. This time depends upon the velocity and the thickness of the highly doped n^+ layer.

- A large time is required to remove the plasma because total plasma charge is large as compared to the charge per unit time in the external current. At point E, plasma is removed. As the residual charge is removed, the voltage increases from point E to point F. At point F, all the charges generated internally has been removed. From point F to point G, the diodes charged up again like a fixed capacitor. At point G, the diode current goes to for half period and the voltage remains constant VA until the current comes back and the cycle repeats.

2.9.5 Advantages

1. It offers higher efficiency of about 15 to 40% as compared a IMPATT diode.
2. It has very low power dissipation.
3. It is most suitable for pulsed operation.
4. It can operate at higher frequencies, about 3 to 50 GHz.

2.9.6 Disadvantages

1. It is not used in continuous operation mode as it offers high power densities, i.e. 10 to 100 W/m^2.
2. It has very high noise figures which is about 60 dB.
3. It supports frequencies below millimeter band.

2.9.7 Applications

1. Microwave beacons.
2. Local oscillator in radars.
3. Instrument landing system.
4. Band pulsed transmitters.
5. Pulsed array radar.
6. Radio altimeter
7. Power amplifiers.
8. Oscillators.

2.10 SCHOTTKY BARRIER DIODE

2.10.1 Introduction

- The Schottky Barrier Diode (SBD) or Schottky diode is another type of semiconductor diode but it has the advantage that their forward voltage drop is substantially less than that of the conventional silicon p-n junction diode. This diode is named after the German physicst Walter H. Schottky.
- Schottky diodes has many useful applications from rectification, signal conditioning and switching through TTL and CMOS logic gates due to mainly to their low power and fast switching speeds. TTL Schottky logic gates are identified by the letter LS appearing somewhere in their logic gate circuit code e.g. 74LS00.

2.10.2 Definition

- The Schottky barrier diode is abbreviated as SBD and it is also known as hot-carrier diode or simply a Schottky diode. The cat's whisker detectors used in the early days of wireless and metal rectifiers, used in early power applications, can be considered primitive Schottky diodes.
- **The Schottky barrier diode (SBD) is a semiconductor diode formed by the junction of a semiconductor with a metal, having a low forward voltage drop and a very fast switching action.**
- When sufficient forward voltage is applied, a current flows in the forward direction. A silicon diode has a typical forward voltage of 600 - 700 mV, while the Schottky's forward voltage is 150 - 450 mV. This lowers forward voltage repairment across higher switching speeds and better system efficiency.

2.10.3 Construction

- **A metal semiconductor junction** is formed between a metal and a semiconductor, creating a **Schottky barriers**, instead of semiconductor junction, as in conventional p-n junction diode. Typical metals used are molybdenum, platinum, chromium or tungsten, and certain silicides, e.g. palladium silicate and platinum silicide, whereas the semiconductors would typically be n-type silicon.

- The metal side acts as the **cathode** of the **anode**, and n-type semiconductor type acts as the **cathode** of the diode, meaning **conventional current** can flow from the metal side to the semiconductor side, but not in the opposite direction. This Schottky barrier results in both very fast switching and slow forward voltage drop.
- The choice of the combination of the metal and semiconductor determines the forward voltage of the diode. Both p- and n-type semiconductors can develop Schottky barriers. However, the p-type typically has a much lower forward voltage of the diode. As the reverse leakage current is increased dramatically with lowering the forward voltage. It cannot be too low, so the usually employed range is about 0.5 to 0.7 V, the p-type, semiconductors are employed rarely.
- **Titanium silicide** and other refractory silicides, which are able to withstand the temperatures needed for source/drain annealing in CMOS processes, usually have too low forward voltage to be useful. So, processes using these silicides therefore usually do not offer Schottky diodes.
- With increased doping of the semiconductor the width of the depletion region drops. Below certain width, the charge carriers can tunnel through the depletion region. At very high doping levels, the junction does not behave as a rectifier any more and becomes an ohmic contact. This can be used for the simultaneous formation of ohmic contacts and diodes, as a diode will be formed between the silicide and the heavily doped p-type or n-type region.
- Lightly doped p-type regions pose a problem, as the resulting contact has too high a resistance for a good ohmic contact, but too low formed voltage and too high reverse leakage to make a good diode.
- As the edge of the Schottky contacts are fairly sharp a high electric field gradient occurs around them, which limits how large the reverse breakdown voltage threshold can be. Various strategies are used, from guard rings to overlaps of metallization to spread out the field gradient.
- The guard rings consume variable die area and are used primarily for larger higher voltage diodes, while overlapping metallization is employed primarily with smaller low-voltage diodes. Schottky diodes are often used in anti-saturation clamps in **Schottky transistors**. Schottky diodes made from palladium (PmSi) are excellent due to their lower forward voltage, which have to be lower than the forward voltage of the base-collector (B-C) junction.
- The Schottky temperature coefficient is lower than the coefficient of the B-C junction, which limits the use of PdSi at higher temperature. For power Schottky diodes, the parasitic resistances of the buried n^+ layer and the epitaxial n-type layer becomes important. The resistance of the epitaxial layer is more important than it is for a transistor, as the channel must cross its entire thickness.
- However, it serves as a distributed ballasting resistor over the entire area of the junction, and under usual conditions prevents localized thermal runaway.

2.10.4 Features

1. Higher efficiency.
2. Low forward voltage drop.
3. Low capacitance.
4. Low profile, surface mount package, ultra small.
5. Integrated guard ring for stress protection.

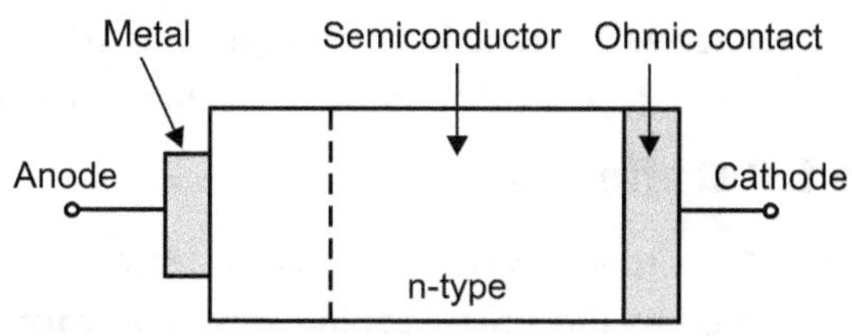

Fig. 2.18 : Physical structure of Schottky diode

2.10.5 Advantages

1. Low turn-on voltage.
2. Free recovery time.
3. Low junction capacitance.
4. Higher efficiency.
5. Low forward voltage drop.
6. Fast recovery time.

2.10.6 Limitations

1. High several leakage current.
2. Thermal instability.

2.10.7 Applications

1. Voltage cloning
2. Prevention of transistor saturation.
3. Standalone photovoltaic system.
4. Solar cell panels.
5. Rectifiers.
6. RF mixer.
7. Detector mixers.
8. Power or circuits.

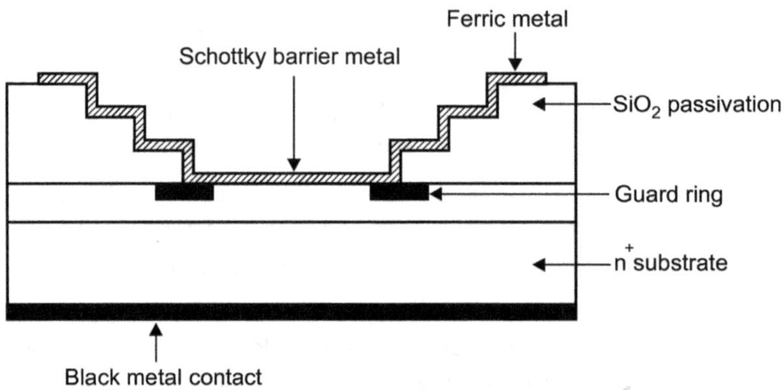

Fig. 2.19 : Schottky barrier diode

2.11 MASER

2.11.1 Brief History

- In 1952, the theoretical principles governing the operation of a MASER were first described by **Joseph Weber** in the Electron Tube Research Conference in **Otavia**. Independently, **Charles Hard Townes, Jymes P. Gordon** and **H. J. Zieger** built the first ammonia Maser at Columbia University in 1953. This device used stimulated emission in a stream of energized **ammonia** molecules to provide amplification of microwaves at a frequency of about 24 GHz.
- **Charles Towens** later worked with **Arthur P. Schawlo** to describe the principle of the Optical Maser or Laser, of which Theodosette Maiman created the first working model in 1960.

2.11.2 Definition

- An acronym MASER stands for Microwave Amplification by Stimulated Emission of Radiaion. Masers are used as the **time keeping device in atomic clocks**, and as extremely low noise microwave amplifiers in **radio telescopes** and deep communication ground (or Earth) stations.
- **A Maser is a microwave device that produces coherent electromagnetic waves through amplification by stimulated emission.**
- Modern masers can be designed to generate electromagnetic waves at not only microwave frequencies, but also radio and infrared frequencies. For this reason, **Charles Townes** suggested replacing **'microwave'** with the word **'molecular'** as the first word in the acronym of Maser.

2.11.3 Construction

- During the early 1960s, the jet propulsion laboratory developed a Maser to provide ultra-low-noise amplification of S-band microwave signals received from deep space probes. This Maser is used deeply refrigerated helium (He) to chill the amplifier down to a temperature of 4° Kelvin.
- Amplification was achieved by exciting a ruby comb with a 12 GHz Klystron. In the early years, it took days to chill and remove the impurities from the hydrogen lines. Refrigeration was a two stage process with a large unit on the ground, and a crosshead compressor within the antenna.

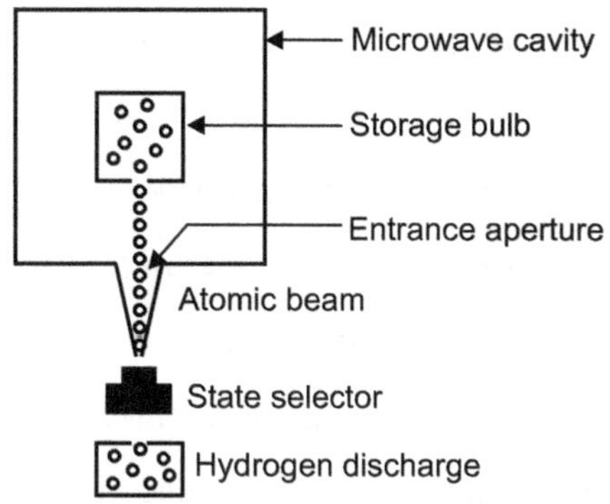

Fig. 2.20 : Hydrogen Maser

- To find injection was at 21 MPa (= 3000 psi) through a 150 μm (= 0.006 inch) micrometer adjustable entry to the chamber. The whole system noise temperature looking at cold sky (2.7 Kelvin in the microwave band) was 17 Kelvins. This gave such a low noise figure that the **Mariner IV space probe** to the earth even though the output power of its radio transmitter was only 15 watts, and hence the total signal power received was only 169 dB with respect to a milliwatt (mW).

2.11.4 Brief History

- A beam of atomic hydrogen is produced. This is done by submitting the hydrogen gas at low pressure to a high-frequency **radio wave** discharge as shown in Fig. 2.20. Then in order to get some **stimulated emission** it is necessary to create a **population inversion** of atoms. This is done in a way that is very similar to the **Stern-Gerlach experiment.**
- After passing through an aperture and a magnetic field, many of the atoms in the beam are left in the upper energy level of the lasing transition. From this state, the atoms (decay) to the lower state and emit some microwave radiation.
- A high quality (Q) factor **microwave cavity** confines the microwaves and reinjects them repeatedly into the atom beam. The stimulated emission amplifiers the microwaves on each pass through the beam. This combination of **amplification** and **feedback** is what defines all **oscillators.**
- The **resonant frequency** of the microwave cavity is tuned to the frequency of the hyperfine energy transition of hydrogen. A small fraction of the signal in the microwave cavity is coupled into a coaxial cable and then sent to a coherent **radio receiver**. The microwave signal coming out the Maser is very weak (a few picowatt). The frequency of the signal is fixed and extremely stable. The coherent receiver is used to amplify the signal and change the frequency. This is done using a series of **Phase-Locked Loops** (PLL) and a high performance **quartz oscillator.**

2.11.5 Applications

1. Low noise amplifier in radio telescopes.
2. Atomic frequency standard.
3. Microwave amplifiers.

Practice Questions

1. What are the limitations of combinational solid state devices ?
2. What is microwave BJT ?
3. What are the requirements of microwave BJT ?
4. Describe the construction of microwave BJT.
5. List applications of of microwave BJT.
6. What is microwave FET ?
7. Describe construction microwave FET.
8. State advantages of microwave FET.
9. List important applications of microwave FET.
10. What is Schottky contact ?
11. Draw the construction and explain working principle of a PIN diode.
12. Describe Gunn diode with its neat diagram.
13. State applications of Gunn diode.
14. With a neat diagram, explain construction, working principle of an IMPATT diode.
15. State applications of an IMPATT diode.
16. What is a Gunn diode ?
17. Explain the tunneling phenomenon.
18. Explain the working principle of a tunnel diode.
19. State applications of a tunnel diode.
20. How tunnel diode can be used as an oscillator ?
21. How tunnel diode can be used as an amplifier ?

✍ ✍ ✍

Chapter **3**

MICROWAVE COMPONENTS

Syllabus

Analysis of Microwave components using s-parameters, Junctions (E, H, Hybrid), Directional coupler, Bends and Corners, Microwave posts, S.S. tuners, Attenuators, Phase shifter, Ferrite devices (Isolator, Circulator, Gyrator), Cavity resonator, Matched termination.

3.1 SCATTERING (s) PARAMETERS

3.1.1 Background

- The first published description of **scattering (s) parameters** was in the thesis of **Vitold Belevitch** in 1945. The name used by **Vitold Belevitch** was **repartition of matrix** and limited consideration to lumped-element networks. The term **scattering matrix** was used by physicist and engineer **Raber Henry Dicke** in 1947, who independently developed the idea during war time work on radar.
- In these s-parameters and scattering matrices the scattered waves are so called travelling waves. A different kind of s-parameters was introduced in 1960s. The later was popularized by **Keneyuki Kurokawa**, who referred to the new scattered waves as **'power waves'**. The two types of s-parameters have very different properties and must not be mixed up.
- Keneyuki Kurokawa in his research paper clearly distinguished the power wave s-parameters and the conventional, tracing wave s-parameters. A variant of the latter is the pseudo-travelling wave s-parameters. In the s-parameter approach, the electric network is regarded as a **'black box'** containing various inter-connected basic electrical circuit components or **lumped elements,** such as resistors, capacitors, inductors and transistors, which interact with other circuits through **ports**.
- The network is characterized by a square matrix of complex numbers called its s-parameter matrix, which can be used to calculate its response to signals applied to the ports.
- Modern vector network analysis measure amplitude and phase of voltage travelling wave **phasors** using essentially the same circuit as that used for the demodulation of **digitally modulated** wireless signals. Many electrical properties of networks of passive components may be using s-parameters, such as gain, return loss, Voltage Standing Wave Ratio (VSWR), reflection coefficient and amplifier stability.

3.1.2 Definition

- Scattering parameters are called, in short, s-parameters or s-matrix. **The parameter or a matrix, which describes the behaviour of linear electrical networks, when undergoing various steady state stimuli by electrical signals.**

3.1.3 Requirements

The following information must be defined, when specifying a set of s-parameters.

1. Frequency.
2. Nominal characteristic impedances.
3. Allocation of port numbers.
4. Conditions which may affect the networks such as temperature, control voltage, and bias current.

3.2 ANALYSIS OF MICROWAVE COMPONENTS

3.2.1 Introduction

- A microwave network is formed, when several microwave devices or components are connected together by transmission lines for transmission of microwave signals.
- **A general microwave one port network is defined by a device or component for which power can enter and leave from only one transmission line or a waveguide.**

3.2.2 Analysis

- We assume that one port network is defined by a surface which is perfectly conducting except for an opening at the terminal connection from where the input signal comes to the microwave devices or components also known as port as shown in Fig. 3.1 (a).

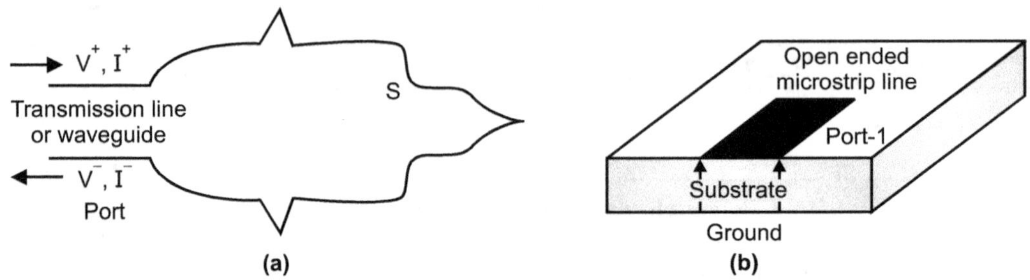

Fig. 3.1 : Microwave engineering

- In Fig. 3.1 (b), an example of 1-port network microwave device is shown. It is basically an open ended microstrip line with ports as input port.
- Fig. 3.1 (a) is general representation of 1-port network microwave device as shown in Fig. 3.1 (b). It is an example of 1-port microwave device. Two port networks are usually used to represent a microwave device. A general 2-port microwave is shown in Fig. 3.2 (a), (b) and (c) respectively. The impedance and admittance matrix representation of this 2-port microwave network is shown below.

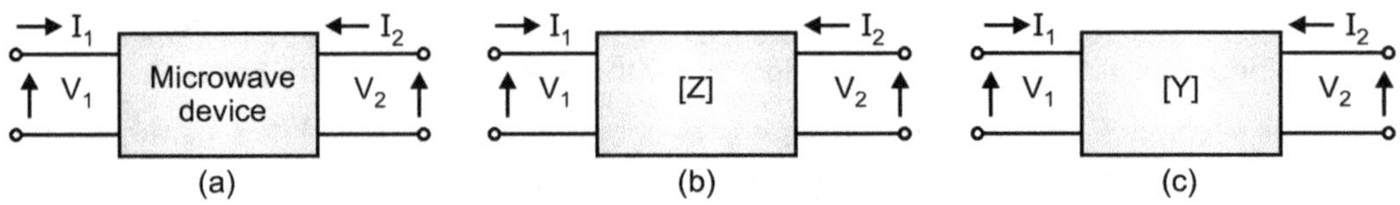

Fig. 3.2 : 2-port microwave network

- Fig. 3.2 (a) is a 2-port microwave device and Fig. 3.2 (b) and (c) show Z-matrix and Y-matrix representation.
- For this case, we can simply write the relation between the port currents and voltages as follows :

$$V_1' = Z_{11}I_1 + Z_2 I_2 \quad \text{... (3.1)}$$

$$V_2' = Z_{21}I_1 + Z_{22}I_2 \quad \text{... (3.2)}$$

- In matrix form, port voltages are expressed in terms of port currents as follows :

$$\begin{bmatrix} V_1 \\ V_2 \end{bmatrix} = \begin{bmatrix} Z_{11} & Z_{12} \\ Z_{21} & Z_{22} \end{bmatrix} \begin{bmatrix} I_1 \\ I_2 \end{bmatrix} \text{ for a microwave device Z-port} \quad \text{... (3.3)}$$

$$\begin{bmatrix} V_1 \\ \vdots \\ V_N \end{bmatrix} = \begin{bmatrix} Z_{11} & Z_{1N} \\ \vdots & \vdots \\ Z_{N1} & Z_{NN} \end{bmatrix} \begin{bmatrix} I_1 \\ \vdots \\ I_2 \end{bmatrix} \text{ for on N-port microwave device} \quad \text{... (3.4)}$$

where Z_{ij} is defined as $\left.\frac{I_i}{V_j}\right|_{I_K = 0,\ K = j}$ for any N-port microwave device.

- Similarly, for admittance matrix representation for an N-port microwave device is given by

$$\begin{bmatrix} I_1 \\ \vdots \\ I_N \end{bmatrix} = \begin{bmatrix} Y_{11} & Y_{1N} \\ \vdots & \vdots \\ Y_{N1} & Y_{NN} \end{bmatrix} \begin{bmatrix} V_1 \\ \vdots \\ V_2 \end{bmatrix} \quad \text{... (3.5)}$$

where Y_{ij} to defined as $\left.\frac{I_i}{V_j}\right|_{V_K = 0,\, K = j}$ ... (3.6)

- In transmission matrix representation also called as [ABCD] matrix, we try to get the voltages and currents of first part in terms of second port voltages and currents of the network.
- Fig. 3.3 (a) shows a 2-port microwave device trend and Fig. 3.3 (b) shows its transmission matrix representation.

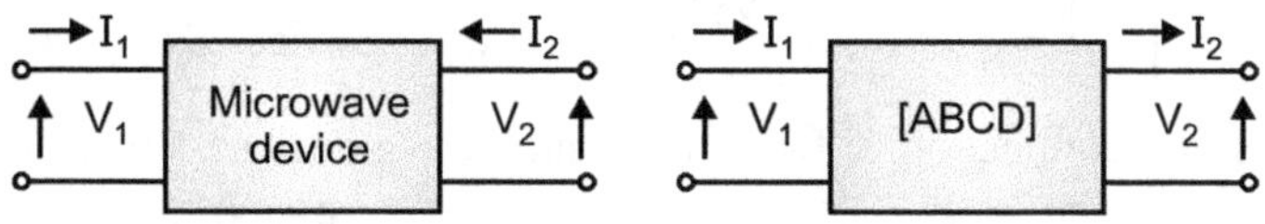

(a) (b)

Fig. 3.3

- Note that the current direction convention of I_2 is opposite to that of the previous 2-port networks to get positive values of B and D parameters of the transmission matrix. The relation between port-1 and port-2 voltages and currents are as follows :

$$V_1 = AV_2 + BI_2$$

$$I_1 = CV_2 + DI_2$$

$$\left.\frac{V_1}{V_2}\right|_{I_2 = 0} \qquad \left.\frac{V_1}{I_2}\right|_{V_2 = 0} \quad \text{... (3.7)}$$

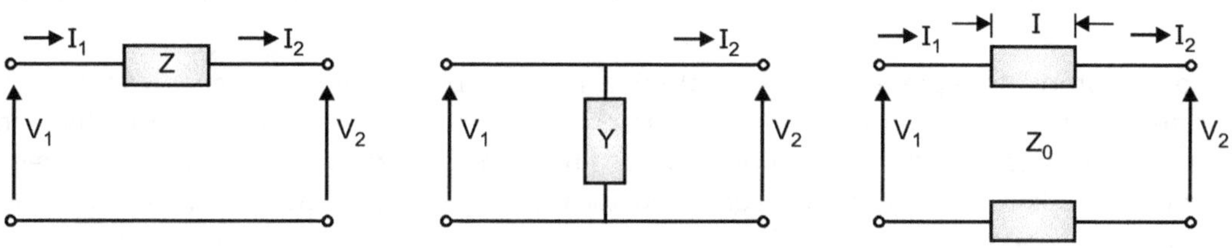

Fig. 3.4

- Series impedance is given by,

$$V_1 = I_2Z + V_2 = V_2 + I_2Z \quad \text{... (3.8)}$$

$$I_1 = I_2 = 0 \times Y_2 + I_2 \quad \text{... (3.9)}$$

$$\therefore \quad \begin{bmatrix} A & B \\ C & D \end{bmatrix} = \begin{bmatrix} 1 & Z \\ 0 & 1 \end{bmatrix}$$

- Shunt impedance is given by,

$$V_1 = V_2 \quad \text{... (3.10)}$$

$$I_1 = V_2Y + I_2 \quad \text{... (3.11)}$$

$$\therefore \quad \begin{bmatrix} A & B \\ C & D \end{bmatrix} = \begin{bmatrix} 1 & 0 \\ 1 & 1 \end{bmatrix} \quad \text{... (3.12)}$$

- Lossless transmission lines are given by

Noting that $\theta = \frac{2\pi l}{\lambda}$ ... (3.13)

$V_1 = V_2\, e^{i\theta} = V_2 \cos\theta + jI_2 \sin\theta = V_2 \cos\theta + j_2I_2Z_0 \sin\theta$ ... (3.14)

$I_1 = I_2\, e^{i\theta} = I_2 \cos\theta + jI_2 \sin\theta = I_2 \cos\theta + jV_2Y_0 \sin\theta$ ... (3.15)

$$\therefore \quad \begin{bmatrix} A & B \\ C & D \end{bmatrix} = \begin{bmatrix} \cos\theta & jZ_0 \sin\theta \\ jY_0 \sin\theta & \cos\theta \end{bmatrix} \quad \ldots (3.16)$$

- For lossy case, we can simply write as under :

$$\begin{bmatrix} A & B \\ C & D \end{bmatrix} = \begin{bmatrix} \cosh\gamma l & Z_0 \sin\gamma l \\ Y_0 \sinh\gamma l & \cosh\gamma l \end{bmatrix} \quad \ldots (3.17)$$

- In practice, many networks consist of a series connection of two or more 2-port networks. We will assume that the overall 2-port ABCD matrix of two or more 2-port networks can be found by simply multiplying the 2-port networks of the individual 2-port networks.

$$\begin{bmatrix} V_1 \\ I_1 \end{bmatrix} = \begin{bmatrix} A_1 & B_1 \\ C_1 & D_1 \end{bmatrix} \begin{bmatrix} V_2 \\ I_2 \end{bmatrix} \text{ for microwave device - 1} \quad \ldots (3.18)$$

$$\begin{bmatrix} V_2 \\ I_2 \end{bmatrix} = \begin{bmatrix} A_2 & B_2 \\ C_2 & D_2 \end{bmatrix} \begin{bmatrix} V_3 \\ I_3 \end{bmatrix} \text{ for microwave device - 2.} \quad \ldots (3.19)$$

- Fig. 3.5 shows the series connection of two 2-port microwave devices. Hence, it is given by,

$$\begin{bmatrix} V_1 \\ I_1 \end{bmatrix} = \begin{bmatrix} A_1 & B_1 \\ C_2 & D_1 \end{bmatrix} \begin{bmatrix} A_2 & B_2 \\ C_2 & D_2 \end{bmatrix} \begin{bmatrix} V_3 \\ I_3 \end{bmatrix} = \begin{bmatrix} A & B \\ C & D \end{bmatrix} \begin{bmatrix} V_3 \\ I_3 \end{bmatrix} \quad \ldots (3.20)$$

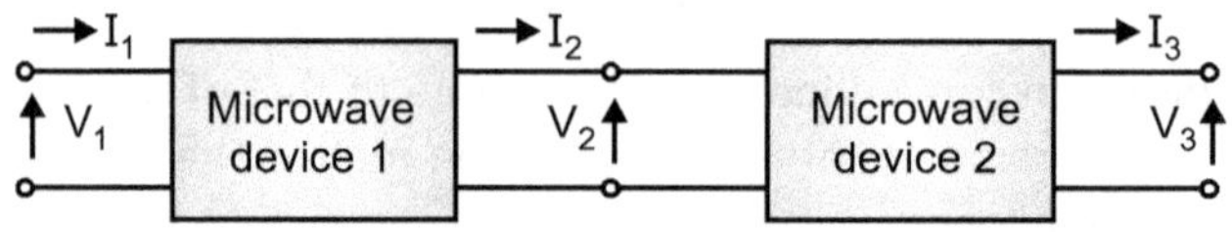

Fig. 3.5

- It is very convenient way of representing the microwave networks.

1. Scattering Matrix Representations :

- Practical problem exists, when trying to measure voltages and currents at microwave frequencies. Direct measurements cannot be done, since all are electromagnetic (EM) waves at high frequencies. It is an abstract idea.
- A more appropriate representation in scattering matrix representation, since we can directly measure reflected, transmitted and incident waves. Scattering matrix relates the incident, transmitted and reflected waves at the port 0. Scattering matrix can be directly measured, with a network analyzer. Once the parameter are known, they can be converted to any other matrix parameters for the analysis of the microwave networks.

2. Two-Port Network Analysis :

- If we denote the amplitude of incident voltage wave at port-1 and the amplitude of reflected voltage wave from port-1, then scattering parameters are expressed as

$$V_1^- = S_{11}V_1^+ + S_{12}V_2^+ \quad \ldots (3.21)$$

$$V_2^- = S_{21}V_1^+ + S_{22}V_2^+ \quad \ldots (3.22)$$

where V_2^- and V_2^+ are respectively the amplitudes of incident and reflected voltage wave at the port-2. In matrix form it is given by,

$$\begin{bmatrix} V_1^- \\ V_2^- \end{bmatrix} = \begin{bmatrix} S_{11} & S_{12} \\ S_{11} & S_{22} \end{bmatrix} \begin{bmatrix} V_1^+ \\ V_2^+ \end{bmatrix} \quad \ldots (3.23)$$

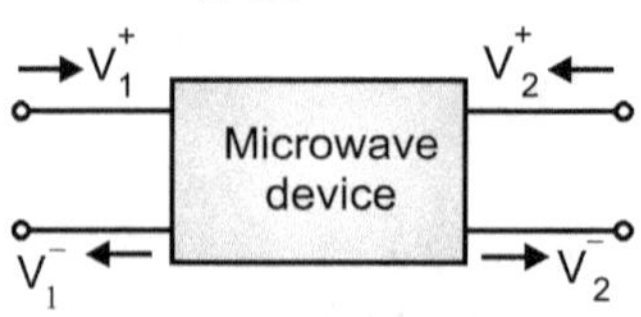

Fig. 3.6

- Fig. 3.6 shows incident and reflected voltage waves in a 2-port network.
- For N-port network :

$$\begin{bmatrix} V_1^- \\ V_2^- \\ \vdots \\ V_N^- \end{bmatrix} = \begin{bmatrix} S_{11} & S_{12} \ldots & S_{1N} \\ S_{21} & S_{22} \ldots & S_{2N} \\ \vdots & & \\ S_{N1} & S_{N2} \cdots & S_{NN} \end{bmatrix} \begin{bmatrix} V_1^+ \\ V_2^+ \\ \vdots \\ V_N^+ \end{bmatrix} \quad \ldots (3.24)$$

where S_{ij} is defined as $\left.\frac{V_1^-}{V_j^+}\right|_{V_K = 0}$, K^{-1} j

- It means that it is the ratio of amplitude of voltage wave reflected at the port i, when a voltage wave of amplitude V_0^+ is incident at the port j.

3. Reciprocal Microwave Network :

- **If the port 1 and port 2 are interchanged for a 2-port network and the performance of the microwave device is still the same, then we call that network is reciprocal.** Reciprocal [Z] matrix for a reciprocal network is symmetric. Correspondingly, since $[Z] = [Y]^{-1}$, [Y] matrix is also symmetric, i.e. $[Z] = [Z]^1$ and $[Y] = [Y]^1$ for a reciprocal network.
- To do this, let us find the relation between [Z] and [S] matrix. Let us assume that all the ports have same characteristic impedance, i.e.,

$$Z_{01} = Z_{02} = Z_{03} = \ldots = Z_{0N} = Z_0 \quad \ldots (3.25)$$

For an nth port of the microwave network,

$$V_n = I_n^+ + V_n^- \quad \ldots (3.26)$$

$$I_n = I_n^+ - I_n^- \quad \ldots (3.27)$$

$$= \frac{V_n^+}{Z_0} - \frac{V_n^-}{Z_0} \quad \ldots (3.28)$$

- In matrix form, it is given by,

$$V = [V^+] + [V^-] \quad \ldots (3.29)$$

$$= [Z] \cdot [I] \quad \ldots (3.30)$$

$$= [Z] ([I^+] - [I^-]) \quad \ldots (3.31)$$

$$= [Z] \frac{1}{Z_0} ([V^+] - [V^+]) \quad \ldots (3.32)$$

$$\therefore \quad \left([U] + \frac{1}{Z_0}[Z]\right)[V^-] = \left[\frac{1}{Z_0}[Z] - [U]\right] \cdot [V^+] \quad \ldots (3.33)$$

which gives the scattering matrix in terms of Z matrix.

Similarly, $[Z]\,[S] + (Z_0)\,[S] = [Z] - (Z_0)\,[U]$ ... (3.34)

$$\therefore \quad [Z] ([S] - [U]) = - Z_0 [S] - (Z_0) [U] \quad \ldots (3.35)$$

$$\therefore \quad [Z] = (Z_0) ([S] + [U]) (- [U] - [S])^{-1} \quad \ldots (3.36)$$

which gives Z matrix in terms of scattering parameters.

By adding equations (3.26) and (3.28), we get,

$$V_n^+ = \frac{1}{2} (V_n + Z_0 I_n) \quad \ldots (3.37)$$

and by subtracting equations (3.26) from (3.28), we get,

$$V_n^- = \frac{1}{2}(V_n - Z_0 I_n) \qquad \dots (3.38)$$

In matrix form, it is given by,

$$[V^+] = \frac{1}{2}([V] + Z_0[I]) \qquad \dots (3.39)$$

$$= \frac{1}{2}([Z]\,[I] + Z_c[I])$$

$$= \frac{1}{2}([Z] + Z_0[U])\,(I) \qquad \dots (3.40)$$

$$\begin{bmatrix} V^- \\ \hline V^+ \end{bmatrix} = [S] = ([Z] - Z_0[U])\,([Z] + Z_0[U])^{-1} \qquad \dots (3.41)$$

$$\therefore \quad [S]^1 = (([Z] + Z_0[U])^{-1})^1\,([Z] - Z_0[U])^1 \qquad \dots (3.42)$$

$$= ([Z] + Z_0[U])^{-1}\,([Z] - Z_0[U])$$

- Since all the matrix components in RHS of the equation (3.42) are symmetric, hence the scattering matrix is also symmetric.

4. Reciprocal and Lossless Network:

If a network is lossless, then no real power can be delivered to the network.

$$P_{av} = \frac{1}{2} R_e\,([V]^1\,[I]^*) \qquad \dots (3.43)$$

where,

$$[V] = \begin{bmatrix} V_1 \\ V_2 \\ \vdots \\ V_N \end{bmatrix} [I] = \begin{bmatrix} I_1 \\ I_2 \\ \vdots \\ I_N \end{bmatrix}, [Z] = \begin{bmatrix} Z_{11} & Z_{12} & \dots & Z_{1N} \\ Z_{21} & Z_{23} & \dots & Z_{2N} \\ \vdots & & & \\ Z_{N1} & Z_{N2} & \dots & Z_{NN} \end{bmatrix} \qquad \dots (3.44)$$

$$= \frac{1}{R_e}([I]^1 \cdot [Z]^1\,[I]^*) \qquad \dots (3.45)$$

$$= \frac{1}{2} R_e\,\{I_j Z_{11} I_1^* + I_1 Z_{12} I_2^* + I_2 Z_{21} I_1^* + \dots\} \qquad \dots (3.46)$$

$$= \frac{1}{2} R_e \left\{ \sum_{m=1}^{N} \sum_{n=1}^{N} I_m Z_{mn} I_n^* \right\} \qquad \dots (3.47)$$

- Since, $[I_n]_s$ are independent, we would set all port elements equal to zero except at the nth port current. Hence,

$$R_e\,\{I_m Z_{mn} I^*\} = 0 = [I_n]^2\,R_e\,\{Z_{nn}\} = 0 \qquad \dots (3.48)$$

$\therefore \quad R_e\,\{Z_{nn}\} = 0$. Hence, Z_{nn} is purely imaginary

- Now let us set all the port elements to zero, except at the mth and n[th] ports. In that case,

$$R_e\,\{I_m I_{mm} I_n^* + I_n Z_{nm} I_m^* + |\,I_n\,|^2 Z_{nn} + |\,I_m|^2 Z_{mn}\} = 0 \qquad \dots (3.49)$$

For a reciprocal network, since $Z_{nm} = Z_{mn}$

$$\therefore \quad R_e\,\{\{I_m I_{mn} I_n^* + I_n Z_{nm} I_m^*\} = 0 \qquad \dots (3.50)$$

Since, $I_m I_m^* - I_n I_m^*$ is purely a real number.

$\therefore \quad R_e\,\{Z_{nm}\} = 0$

- Hence, the impedance matrix is purely imaginary for a lossless network. Similarly, the admittance matrix is also purely imaginary. The characteristic [S] matrix for a lossless network is given by,

$$P_{av} = \frac{1}{2} R_e ([V]^1 [I]^*) \quad \text{... (3.51)}$$

$$= \frac{1}{2} R_e \{([V'^+]^1 [V^-]^1) \frac{1}{Z_e} ([V^+]^* - [V^-]^*)\} \quad \text{... (3.52)}$$

$$= \frac{1}{2 Z_0} R_e \{([V^+]^1 [V^+]^* - [V^+]^1 [V^-]^* + [V^-]^1 [V^+]^* - [V^-]^1 [V^-]^*)\} \quad \text{... (3.53)}$$

$$= \frac{1}{2Z_0} R_e \{[V^+]^1 [V^+]^* - [V^-]^1 [V^-]^*]\} \quad \text{... (3.54)}$$

where, the first term is the incident power and the second term is the reflected power

1. $[V^-]^1 [V^+]^* - [V^+]^1 \ [V^-]^*$ is purely imaginary.

Proof :

$$[V^-]^1 = [V^- \overset{-}{V}_2 \ldots \overset{-}{V}_N, \quad [V^+]^* = \begin{bmatrix} \overset{+}{V}_1 {}^* \\ \overset{+}{V}_2 {}^* \\ \overset{+}{V}_N {}^* \end{bmatrix} \quad \text{... (3.55)}$$

$$[V^+]^1 = [\overset{+}{V}_1 \overset{+}{V}_2 \ldots \overset{+}{V}_N]$$

$$\therefore \quad [V^-]^1 [V^+]^* - [V^+]^1 [V^+]^* [V^-]^* = \begin{bmatrix} \overset{-}{V}_1 {}^* \\ \overset{-}{V}_2 {}^* \\ \overset{-}{V}_N {}^* \end{bmatrix} \quad \text{... (3.56)}$$

$$= \overset{-}{V}_1 \overset{+}{V}_1 + \overset{-}{V}_1 \overset{+}{V}_2 + \ldots - \overset{-}{V}_N \overset{+}{V}_N$$

$$= (\overset{-}{V}_1 \overset{+}{V}_1{}^* - \overset{+}{V}_1 + \overset{-}{V}_1{}^*) + (V_2 \overset{+}{V}_2{}^* - \overset{+}{V}_2 \overset{-}{V}_2{}^*) + (\overset{-}{V}_N \overset{-}{V}_N{}^* - \overset{+}{V}_N \overset{-}{V}_N{}^*)$$

All the terms are purely imaginary.

2. $[V^+]^1 [V^+]^*$ and $[V^-] [V^-]^*$ are purely real.

$$[V^+] [V^+]^* = \overset{+}{V}_1 \overset{+*}{V}_1 + \overset{+}{V}_2 \overset{+}{V}_2 + \ldots + \overset{+}{V}_N \overset{+*}{V}_N$$

$$= [\overset{+}{V}_1]^2 + |\overset{+}{V}_2|^2 + |\overset{+}{V}_3| + \ldots |V_N|^2 \text{ which is a real number.}$$

Similarly, $[V^-]^1 [V^-]^* = [V_1]^2 + |\overset{-}{V}_2|^2 + |\overset{-}{V}_3|^2 + \ldots |\overset{-}{V}_N|^2$ which is real number.

- For a lossless network, $P_{av} = 0$

$$\therefore \quad [V^+]^1 [\overset{+}{V}_1]^* = [V^-]^1 [V^-]^*$$

$$[V^+]^1 [V^+]^* = [V^+]^1 [S]^1 [V^-]^* \Rightarrow \sum_{k=1}^{N} S_{Ki} S_{Kj} = S_{ij} \quad \text{... (3.57)}$$

$$= [V^+]^1 [S]^1 [S]^* [V^+]^8$$

$$\therefore \quad [S]^1 [S]^* = U$$

[S] is a unitary matrix.

6. Network Matrix Transformation :

(a) From the distribution of ABCD matrix,

$$V_1 = AV_2 + BI_2$$

$$= A\left(\frac{I_1 - NI_2}{C}\right) + BI_2$$

$$= \frac{A}{C} I_1 - \left(\frac{AD - BC}{C}\right) I_2$$

$$= Z_{11}I_1 - Z_{12}I_2 \quad \text{... (3.58)}$$

Note that I_2 direction is opposite for [Z] and [ABCD] matrix representations. Similarly,

$$V_2 = \frac{I_1 - DI_2}{C}$$

$$= \frac{1}{C} I_1 - \frac{D}{C} I_2 = Z_2 I_1 - Z_{22} I_2$$

$$\therefore \quad Z = \begin{bmatrix} Z_{11} & Z_{12} \\ Z_{21} & Z_{22} \end{bmatrix} = \frac{1}{C}\begin{bmatrix} A & AD - BC \\ 1 & D \end{bmatrix} \quad \text{... (3.59)}$$

For symmetrical network, $Z_{21} = Z_{22} \Rightarrow A = D$

For reciprocal network, $Z_2 = Z_{21}; \; \frac{AD - BC}{C} = \frac{1}{C} \Rightarrow AD - BC = 1$

3.3 MULTIPLE JUNCTIONS

3.3.1 Need

- When it is required to combine two or more signals or split a signal into two or more parts in a waveguide system, some form of multiple junction must be used.
- Therefore, there is a need of multiple junction to combine two or more electromagnetic signals or split the electromagnetic signal into two or more parts in a wave guide system.

3.3.2 Definition

- The waveguide junction is an interconnection of two or more microwave components. Waveguide junction is also known as a Tee.
- **Waveguide junctions which are used to connect a branch or section of waveguide in series or parallel with main waveguide transmission line for splitting also of combining power in waveguide is called multiple junction.**
- These multiple junctions have three identical parts. Since, the electric field lines changes their direction, when they come out of Port 1 and Port 3 it is known as E-plane junction. E-plane junction is also known as voltage or series junction, symmetrical about the central arm.

3.3.3 Types

1. E-plane junction.
2. H-plane junction.
3. Hybrid (or) magic junction
4. Rat race ring junction.

3.4 E-PLANE JUNCTION

3.4.1 Definition

At the electric field, lines change their directions, when they come out of ports (1) and (2), it is known as an E-plane.

3.4.2 Construction

- The construction of E-plane junction is as shown in Fig. 3.7.
- A rectangular slot is cut along the broader dimensions of a long waveguide and a side arm is attached to it.
- It consists of 3 ports. Port 1 and 2 are the collinear arms and port 3 is the E-arm.

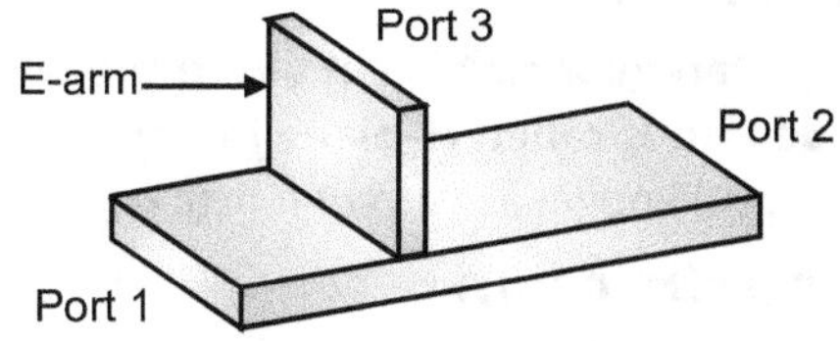

Fig. 3.7 : E-plane junction

3.4.3 Properties

- The properties of E-plane junction can be described by using scattering matrix. The power out of port-3, i.e. side or E-arm is proportional to the difference between instantaneous power entering from port-1 and port-2.
- When power entering the main arm, i.e. port-1 and port-2 are in phase opposition, maximum energy comes out of port-3 or E-arm. It is a 3-port junction, scattering matrix can be derived as follows :
  1. There are three ports. Therefore, [s] is a 3 × 3 matrix.
  2. The scattering coefficients $s_{23} = - s_{13}$. As outputs of port-1 and port-2 are out of phase by 180° with an input at port-3.
  3. If port-3 is perfectly matched in the junction
     $$s_{ij} = 0 \quad \text{... (3.60)}$$
  4. From symmetric property, $s_{ij} = s_{ji}$, $s_{12} = s_{21}$, $s_{12} = s_{31}$ and $s_{23} = s_{32}$
     $$\therefore \quad [s] \begin{bmatrix} s_{11} & s_{12} & s_{13} \\ s_{21} & s_{22} & s_{23} \\ s_{31} & s_{32} & s_{33} \end{bmatrix} \quad \text{... (3.61)}$$
  5. From unitary property, $[s]\,[s] = [1]$ ... (3.62)

3.4.4 Working Principle

- When input is applied to the port 3, the two outputs at port 1 and port 2 will have a phase shift of 180° as shown in Fig. 3.8.
- Signals to be split or two signals that are to be combined will a fed from E-arm.
- Generally, the powers at port 3 is proportional to the difference between instantaneous power enters from power is 1 and 2.

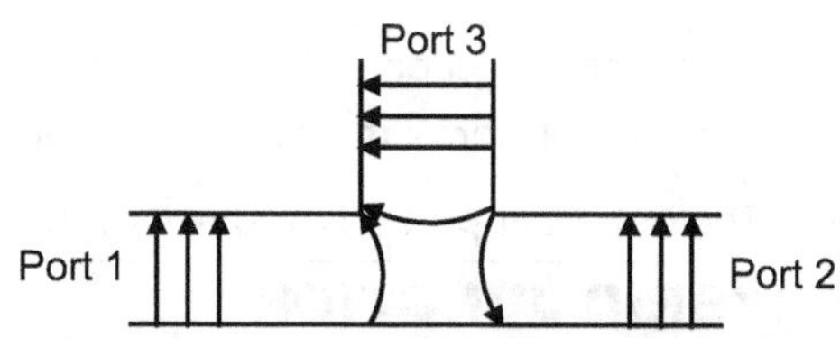

Fig. 3.8 : E-plane junction with electric field

- When power entering the main arm ports 1 and 2 are in phase opposition and maximum energy comes out of port 3 or E-arm.
- It is a three port junction, so that the scattering matrix parameters can be derived. An input at port 3 equally divides between port 1 and 2, but introduces a phase shift of 180° between the two output ports. Thus, E-plane junction also act is 3 dB splitter. If inputs at ports 1 and 2 are equal, then no output at port 3. For E-plane junction, the scattering parameter s is given by,

$$[s] = \begin{bmatrix} \frac{1}{2} & \frac{1}{2} & \frac{1}{\sqrt{2}} \\ \frac{1}{2} & \frac{1}{2} & -\frac{1}{\sqrt{2}} \\ \frac{1}{\sqrt{2}} & -\frac{1}{\sqrt{2}} & 0 \end{bmatrix} \quad \text{... (3.63)}$$

3.5 H-PLANE JUNCTION

3.5.1 Definition

- **A rectangular slot is cut along the width of a main waveguide and attaching another waveguide the side arm is called H-plane junction.**
- It is also known as *current junction.*

3.5.2 Construction

- The construction of H-plane junction is as shown in Fig. 3.9.
- The ports 1 and 2 of the main waveguide are known as collinear ports. The port 3 is called H-arm or side arms.
- In H-plane junction, the axis of side arm is parallel to the planes of the main transmission line.

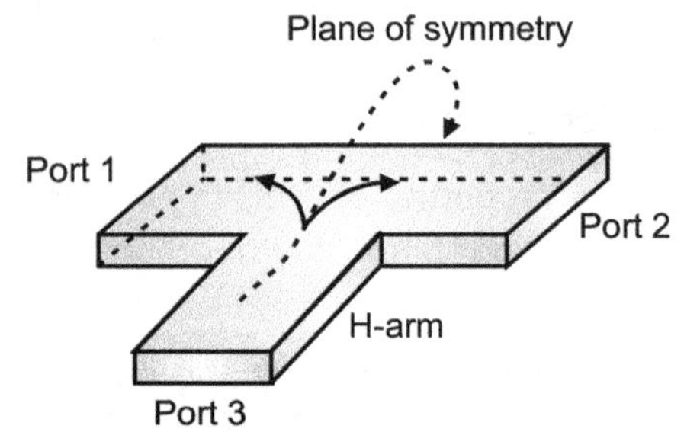

Fig. 3.9 : H-plane junction

3.5.3 Working Principle

- Consider the structure of Fig. 3.9. Since, all three arms of the H-plane junction lie in the plane of magnetic field, the magnetic field divides itself into the arms. Therefore, it is also known as a current junction.
- The properties of H-plane junction are defined by [s] matrix. [s] matrix if scattering parameter s is a 3 × 3 matrix, because of three ports.

$$[s] = \begin{bmatrix} \frac{1}{2} & -\frac{1}{\sqrt{2}} & \frac{1}{\sqrt{2}} \\ -\frac{1}{2} & \frac{1}{2} & \frac{1}{\sqrt{2}} \\ \frac{1}{\sqrt{2}} & \frac{1}{\sqrt{2}} & 0 \end{bmatrix} \quad \dots (3.64)$$

- In H-plane junction, if input is given at port 3 and no inputs at ports 1 and 2, then the power coming out of port 1 or port 2 is 3 dB with respect to input power.
- The output at port 3 is additive of the two inputs at port 1 and port 2 and these are added in phase.

3.6 HYBRID JUNCTION

3.6.1 Definition

- **A hybrid junction is a combination of an E-plane and H-plane junction.** It acts as a 4-port hybrid circuit. It is also known as *hybrid junction.*

3.6.2 Construction

- The construction of a hybrid junction is as shown in Fig. 3.10.
- It consists of four ports, also E and H-arms and also a collinear arm.

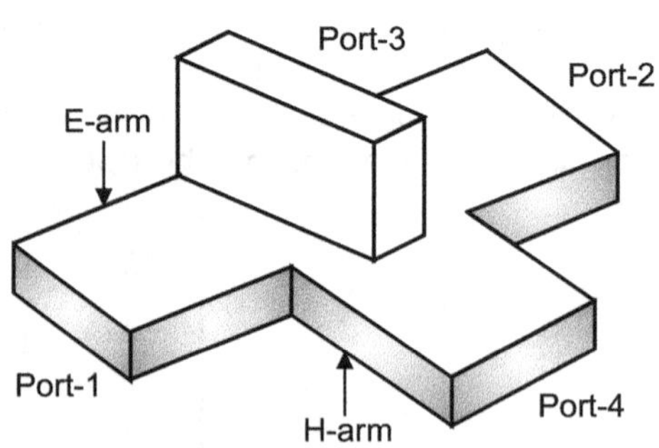

Fig. 3.10 : Hybrid junction

3.6.3 Working Principle

- If two waves of equal magnitude and the same phase are fed into port-1 and port-2, then the output will be zero at port-3 and additive at port-4.
- If a wave is fed into port-4, it will be divided equally between port 1 and port-2 of the collinear arms and will not appear at port-3.

$$s_{14} = s_{41} = \frac{1}{\sqrt{2}}, \; s_{24} = s_{42} = \frac{1}{\sqrt{2}} \text{ and } s_{44} = 0. \quad \text{... (3.65)}$$

- If a wave is fed into port-3, it will produce an output of equal magnitude and opposite phase at port-1 and port-2. The output at port-4 is zero.

$$s_{13} = s_{31} = \frac{1}{\sqrt{2}}, \; s_{24} = s_{42} = \frac{1}{\sqrt{2}} \text{ and } s_{34} = 0$$

- If a wave is fed into one of the collinear arms at port-1 or port-2, it will not appear in the other collinear arm at port-2 or port-1 because the E-arm causes a phase delay, while the H-arm causes a phase lead.

$$\therefore \quad s_{12} = s_{22} = 0 \quad \text{... (3.66)}$$

- Hybrid tee is symmetrical about an imaginary plane bisecting arms port-3 and port-4.
- If port-1 and port-2 are terminated in matched loads and no reflections take place inside the junction, then entrances of power through either port-2 or port-4 result in equal power delivery to arm 1 and arm 2. Reflections may take place due to severe discontinuities in the junction.

3.7 DIRECTIONAL COUPLERS

3.7.1 Introduction

- All four ports of a directional coupler are (ideally) matched and the circuit is (ideally) lossless. Directional couplers can be realized in microstrip, stripline, coax and waveguide. Directional couplers are used for sampling a signal and they generally used for distributed properties of microwave circuits.
- The coupling process generally occurs within a quarter wavelength or multiple quarter wavelength portions of the device. In these distributed couplers, the energy and fields of another distributed structure, coupling signals from one to another.
- These couplers do not rely on the interaction of the fields and waves between transmission structure to create performance similar to coupled line structure, but instead rely on networks composed of independent (uncoupled) circuit elements.
- It is sometimes required to measure power being delivered to a load or an antenna through a transmission line or a waveguide. This is often done by a sampling technique in which a calculated function of the power is measured, so that the total may be calculated.
- In this technique, only forward going wave is measured and the measurement does not take account of the reflected wave, if any. A number of units are available for this purpose and are called directional couplers.

3.7.2 Definition

- **The directional coupler is a four-port junction that is used in wide variety of microwave systems to satisfy almost any requirement for sampling incident and reflected microwave power conveniently and accurately with minimal disturbance to transmission line.**
- The directional coupler is a four port circuit where one port is isolated from the input port. The directional coupler is a passive reciprocal network.
- A directional coupler is an electronic component having four-port circuits with one port being isolated from the input port. A directional coupler is a device that samples a small amount of microwave power for measurement purposes. The power measurement includes incident power, reflected power, VSWR values, etc.

3.7.3 Basic Principle

- The directional coupler is as shown in Fig. 3.11.
- It consists of four ports.
- When a wave travels from port 1 to port 2, a fixed fraction of this power appears at port 4 and there is no output at port 3.

Fig. 3.11 : Directional coupler

- Conversely, if the wave is traveling from port 2 to port 1, a fraction of this signal appears at port 3 and there is no output at port 4.
- When the transmission takes place from port 1 to port 2, the port 4 is called a coupled port and the port 3 is called isolator port.

3.7.4 Types

1. Co-axial unidirectional coupler.
2. Waveguide directional coupler.
3. Two-hole directional coupler.
4. Long slot directional coupler.
5. Bethe hole directional coupler.

3.7.5 Properties

1. All the terminals are matched to the ports.
2. When the power travels from port-1 to port-2, some portion of it gets coupled to port-4, but not to port-3.
3. As it is also a bidirectional coupler, when the power travels from port-2 to port-1, some portion of it gets coupled to port-3, but not to port-4.
4. If the power is incident through port-4, a portion of it is coupled to port-1, but not to port-2.
5. Port-1 and port-3 are decoupled as for port-2 and port-4.

3.7.6 Applications

1. It is used to couple the microwave power, which may be unidirectional or bidirectional.
2. It is used for measurement of microwave power, which includes incident power, reflected power, VSWR values, etc.
3. It can be used to monitor frequency and power level on the signal without interrupting the main power flow in the system.
4. It may be used for coherent power divider which is used in the home to divide cable TV or over the air TV signals to multiple TV sets and other devices.
5. It can be used in a balanced microwave amplifier.

3.8 TWO-HOLE DIRECTIONAL COUPLER

3.8.1 Introduction

- The principle of a multihole directional coupler can be illustrated by considering two holes directional coupler.
- The most common type of the waveguide directional coupler consists of two waveguides placed adjacent to each other such that they share a common wall.

3.8.2 Construction

- Fig. 3.12 shows a typical two-hole waveguide directional coupler. It consists of piece of waveguide to be connected to a piece of auxiliary waveguide.

- It has two holes A and B in a common wall of two waveguides. These holes marked A and B are placed a quarter guide wavelength ($\lambda_o/4$) apart.
- The forward travelling wave couples some energy through each of the two holes from the main waveguides into the auxiliary waveguide.

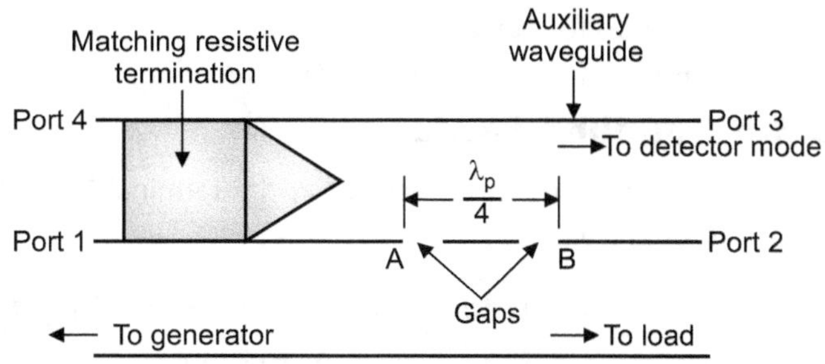

Fig. 3.12 : Two-hole directional coupler

- The signal reaching arm-4 will cover the same distance, irrespective of whether it has travelled clockwise or counterclockwise and so addition will take place at point 4, resulting in some travelling wave down arm-4.
- Similarly, a signal reaching the arm-2 will have travelled a distance of $\lambda_o/4$, if travelling clockwise and a distance of $5\lambda/4$, if travelling counterclockwise. Therefore, two signals will add at that point and propagation will take place down the arm 2.
- Now consider the condition at arm-3. The clockwise signal will travel a distance of $\lambda_p/2$, while the counterclockwise signal will travel a distance of λ_p. This results in cancellation of two signals, or at this point they become out of phase by 180°, so that no signal propagates down the arm 1.
- By similar reasoning, it can be shown that arm is quite similar to that of magic tea, although they have absolutely different principle of operation.
- The rat race and magic tea can be used interchangeably. The rat race is smaller bulk, but it requires internal matching. It does not need much matching, if the thickness of the waveguide used for fabricating the rat race is of correct dimensions. The rat race is preferable at shorter wavelengths, because its dimensions are less critical.
- Also attenuator is used to absorb backward wave components in the auxiliary waveguide. The probes do not actually touch the inner conductor of the auxiliary waveguide. They couple sufficient energy into it simply by being near it.
- If they did not touch, most of the energy in the main waveguide would be coupled into the auxiliary waveguide. The auxiliary is terminated at one end by a resistive load. The other end goes to a detector probe for measurement.

3.8.3 Working Principle

- When one of the waveguide X is excited, the field coupled through each of the holes excites waves in the other waveguide Y. Each of the holes excites two waves, one travelling towards left and the other travelling towards right.
- When the holes are located in the middle of common walls, these two waves have equal amplitude. Because of the quarter wave spacing between two holes, the two waves travelling towards left are out of phase 180° and therefore cancel each other.
- On the other hand, two waves travelling towards right are in phase and thus reinforce each other. Thus, an output appears at port-4 and no power is coupled to the port-3.
- The amount of power coupled to the port-4 depends upon the sizes and the locations of the holes in the common wall. The two-hole directional coupler works satisfactorily only at a single frequency corresponding to quarter wave spacing between the holes.
- The bandwidth of a directional coupler can also both maximally flat and equiripple responses can be designed for. A directional coupler is specified by two factors as under :

  1. The coupling factor. 2. The directivity.

3.8.4 Applications

1. It is used for direct SWR measurements, if the absorbing attenuator is replaced by a detecting device.
2. It is used for measuring components in the auxiliary guide that are proportional to the reflected wave in the main waveguide.

3. It is used for power flow measurement.
4. It is used for unidirectional wave radiation.

3.8.5 Coupling Factor

- Let P_i be the power incident at the terminal 1 and let P_f be the coupled power in the forward direction in terminal 3.
- The coupling factor c is given by, $c = 10 \log \frac{P_i}{P_f} \text{ dB}$ ... (3.67)
- The coupling factor is the ratio of incident power and forward power.

3.8.6 Directivity

- Ideally, the power coupled P_b in backward direction in the arm 4 should be zero. The extent to which this is achieved is measured by the directivity D. The directivity D is given by,

$$D = 10 \log \frac{P_i}{P_b} \text{ dB} \quad \text{... (3.68)}$$

- The directivity is a measure of how well the power can be coupled in the desired direction in the auxiliary line.

3.9 WAVEGUIDE BENDS AND CORNERS

3.9.1 Introduction

- Bends are used, where the change in direction of propagation of the energy in a waveguide is required. There are two types of bends as under :

  (1) H-plane bend, (2) E-plane bend.

3.9.2 Waveguide Bends

- Fig. 3.13 (a) shows a structure of a H-plane bend.
- It is a piece of waveguide smoothly bend in a plane parallel to the magnetic field for the dominant mode. Hence, the name.
- In order to keep the reflections in the bend small, its length is made several waveguides.
- If long length is undesirable because of the size or if a sharp bend is required, then reflections can be minimized by making the mean length of the bend on integral number of guide wavelengths. By doing so, some of the reflections get cancelled.

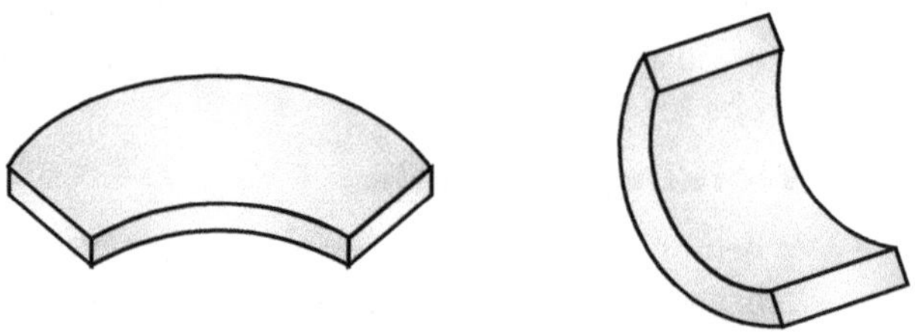

(a) H-plane bend **(b) E-plane bend**

Fig. 3.13 : Waveguide bends

- It should be noted that the sharper the bend, the greater is the mismatch introduced. Fig. 3.13 (b) shows an E-plane bend. At lower frequencies, a bend becomes very inconvenient because of its clumpsy nature.

3.9.3 Waveguide Corners

- For larger wavelengths or lower frequencies, a waveguide bend is rather clumpsy and hence bends become inconvenient, therefore a waveguide corner is a best solution. An ordinary 90° bend would obviously introduce intolerable reflections and to avoid them a part of the corner is cut.
- The dimension depends on wavelength, but if it is correctly chosen, reflections can be almost completely eliminated. The types of waveguide corners are as under :

  1. H-plane motored corner.
  2. E-plane double metered corner.
  3. H-plane double metered corner.

- Fig. 3.14 shows different types of waveguide bends.

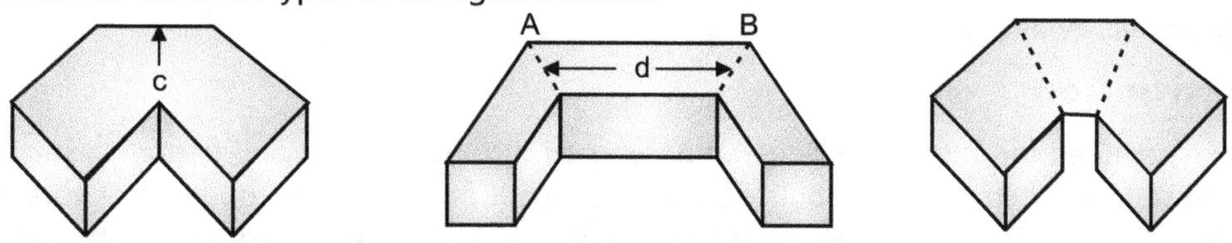

(a) H-plane metered corner **(b) E-plane double metered corner** **(c) H-plane double metered corner**

Fig. 3.14 : Waveguide corners

- The cut corner is said to be metered.
- An E-plane corner is very rarely used, because of the risk of voltage breakdown across the distance 'd' which would be naturally very small in such corner. Thus, if a change of direction of E-plane is desired, then a double metered corner is used as shown in Fig. 3.14 (b).
- In this corner, both the inside and outside walls are cut and the thickness at the cut is same as that of the straight portion of waveguide. If the dimension 'd' is made a quarter of a guide wavelength, reflections from the two cut corners get cancelled, but it makes the corner frequency sensitive.

3.10 FERRITE COMPONENTS

3.10.1 Concept

- The characteristics of wave propagation in ferrite materials are used for designing non-reciprocal components at microwave frequencies. The non-reciprocal components find extensive use in microwave systems.
- Ferrites are ceramic - like materials with a dielectric constant of 10 or more and with specific density that may be as high as 10^{14} times of metals. Properly of Faraday rotation in ferrites is used in designing ferrite components.

3.10.2 Types

- The types of non-reciprocal ferrite components are as under :
  1. Isolators. 2. Circulators. 3. Gyrator.

3.10.3 Faraday's Rotation Principle

- *Faraday's principle* of rotation states that, if a circularly polarized wave (TE_{11}) in a cylindrical waveguide is made to pass through a ferrite rod, which has been influenced by an axial magnetic field B, the axis of polarization gets tilted in the clockwise direction and the amount of tilt depends upon the strength of the magnetic field and geometry of the ferrite.
- This principle is illustrated in Fig. 3.15.

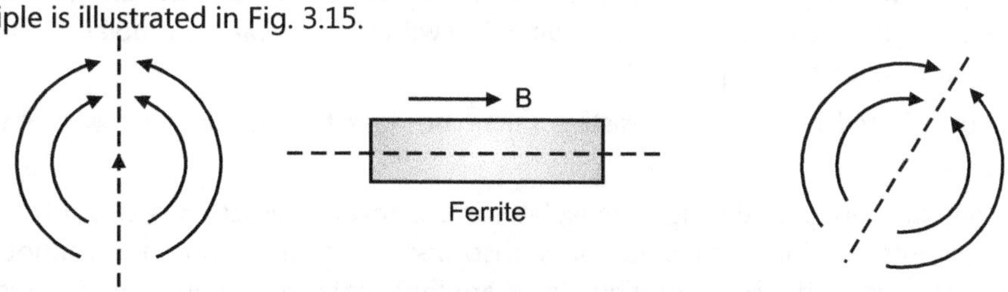

Fig. 3.15 : Faraday's principle of rotation

- This phenomenon is used in microwave components, such as isolator, circulator and gyrator.

3.11 ISOLATORS

3.11.1 Definition

- **An isolator is a two-port device, that transmits microwave or RF power in one direction only.**
- It provides maximum attenuation for transmission from port-2 to port-1. This requirement is very much desirable, when we want to match a source with a variable load. It is used to shield equipment on its input side from the effects of conditions on its output side.

- **An isolator is a non-reciprocal ferrite transmission device, which is used to improve the frequency stability of microwave generators.**

3.11.2 Construction

- Isolation is required for microwave generators, whose output amplitude and frequency may change, when connected to load. The coupling should be unidirectional from generator to load. Isolators made of ferrite materials are used. The operation of ferrite isolator is based on Faraday's rotation or on resonant absorption.
- The Faraday's rotation isolator is shown in Fig. 3.16. It consists of piece of a circular waveguide carrying TE_{11} mode with transition to rectangular waveguide and TE_{10} mode at both ends.

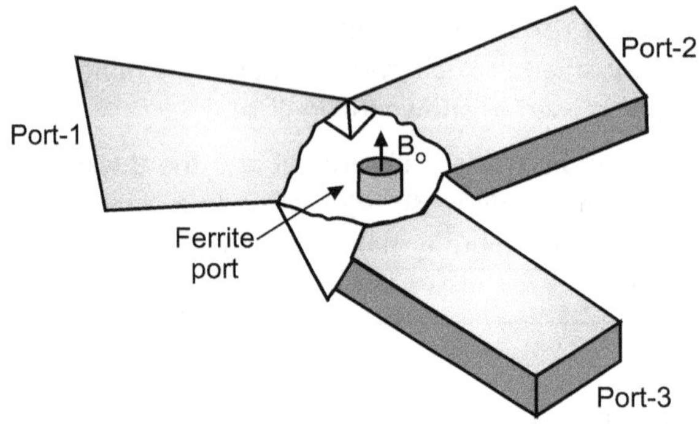

(a)

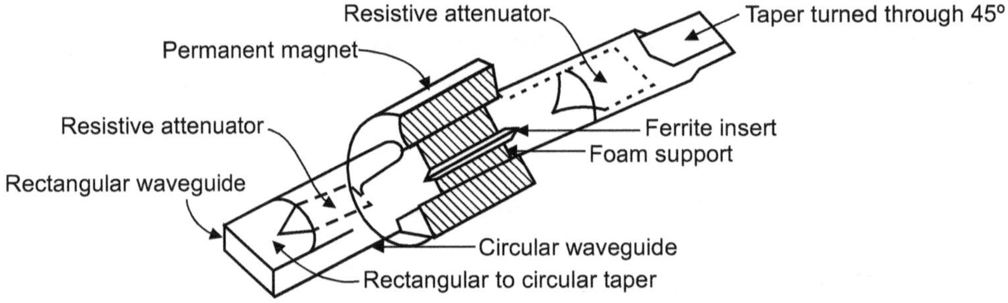

(b)

Fig. 3.16 : Faraday's rotation isolator

- A thin pencil of ferrite is located inside the circular waveguides and waveguide is surrounded by a permanent magnet which generates a magnetic field in ferrite. Because of the d.c. magnetic field applied a wave passing through a ferrite in the forward direction with have its plane of polarites shifted clockwise by 45° by the time it reaches the output end.
- The wave is not affected by either of resistive vanes, because they are at right angles to the plane of its electric field.
- A wave that tries to propagate through the isolator in the reverse direction is also rotated clockwise. Thus, when it emerges into the input transition, it is absorbed by resistive vane and cannot propagate in the input rectangular wave guide. The returning waves are thus highly attenuated and isolation is achieved.
- The resonant absorption isolator is used for high power in the range of 100 W average and 10 kW peak in the X-band. The Faraday's rotation is not suitable for high peak power operation, since non-linear effects occurs at relatively low powers.

3.11.3 Working Principle

- A TE_{10} wave passing from port-1 through the resistive card (attenuator) is not attenuated. After coming out of this card, the wave gets shifted by 45°, because of the twist in anticlockwise direction and then by another 45° in the clockwise direction, because of the ferrite rod.

- Hence, it comes out of port-2 with the same polarization at port-1 without any attenuation. But TE_{10} wave fed from port-2 gets a pass from the resistive card placed near port-2, since the plane of polarization of the wave is perpendicular to the plane of the resistive card.
- Then the wave gets rotated by 45° due to Faraday's rotation in the clockwise direction and further gets rotated by 45° in the clockwise direction due to the twist in the waveguide. Now, the plane of polarization of the wave will be parallel with that of the resistive card and hence the wave will not completely be absorbed by the resistive card and the output at port-1 will be zero.
- This power is dissipated in the resistive card as heat. In practice, 20 to 30 dB isolation is obtained for transmission from port-2 to port-1.

3.11.4 Circulator as an Isolator

- For isolation of transmitters and receivers in a radar system, they are connected to the same antenna. Isolation of input and output in two terminal amplifying device used is a parametric amplifier.
- If one port of a circulator is loaded, it acts as an isolator, i.e. power will pass from port-1 to port-2, but power reflected, back from port-2 will get to the load at port-3 versus going back to port-1. Fig. 3.17 shows implementation of a circulator as an isolator.

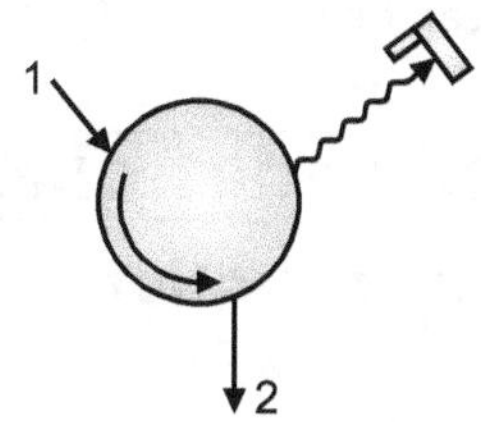

Fig. 3.17 : Circulator as an isolator

3.11.5 Circulator as an Duplexer

- The circulator can be made to work as a duplexer by adding a HPF to port-2.
- Frequencies from port-1 that are below 10 GHz will be reflected by port-2. Frequencies above 10 GHz will pass through port-2.
- Filter could be a piece of waveguide which passes above 10 GHz.
- At 10 GHz cross-over frequency of the duplexer, a 10 GHz signal will be passed to both port-2 and port-1, but will be half power at each port.
- Fig. 3.18 shows implementation of a circulator as a duplexer.

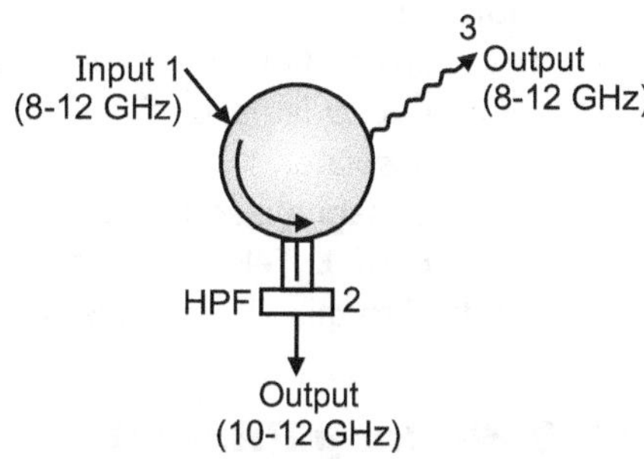

Fig. 3.18 : Circulator as a duplexer

3.12 CIRCULATOR

3.12.1 Definition

- **A circulator is a passive, non-reciprocal three or four port device, in which a microwave or RF signal entering any one port is transmitted to the next port in rotation only.**
- **A port in this context is a point where an external waveguide or transmission line such as a microstrip line or a co-axial cable, connects to the device.**
- A circulator is a multipoint device in which RF signals travel from one port to the next in one direction, either clockwise or anti-clockwise only. The signal to be transmitted travels from port-1 to port-2 and signal received at antenna travels from port-2 to port-3.

3.12.2 Types

- Depending on the materials involved, circulators fall into two main categories as under :

1. **Ferrite circulators :**
   - Ferrite circulators are RF circulators, which are composed of magnetized ferrite materials.
   - They fall into two main classes as under :

     (i) 4-port waveguide circulators. (ii) 3-port Y-junction circulators.

2. Non-ferrite circulators :

- Early work on non-ferrite circulators include active circulators using transistors that are non-reciprocal in nature.
- In contrast to ferrite circulators, which are passive devices, active circulators require power.
- Major issues associated with transistor-based active circulators are the power limitation and the signal noise (S/N) degradation, which are critical when it is used as a duplexer for sustaining the strong transmit power and clean reception of the signal from the antenna. Varactor diode offers one solution.

3.12.3 Applications

1. It is used to shield equipment on its input side from the effects of conditions on its outside side.
2. It can be used as an isolator.
3. It can be used as a type of duplexer in Radar.
4. It can be used as a reflection amplifier, which is a type of microwave amplifier.

3.13 FOUR PORT CIRCULATOR

3.13.1 Introduction

- Methodology used for designing the isolators may be used for the design of a four port circulator. Design of a circulator is based on the principle of *Faraday's rotation*. A circulator is a ferrite component usually having four ports.

3.13.2 Construction

- The construction of a four port circulator is as shown in Fig. 3.19.
- The construction of a four port Faraday's rotation circulator is similar to the Faraday's rotation isolator.
- It uses a magnetized ferrite rod to provide a 45° rotation of E field by using an arrangement identical to that in case of isolator.

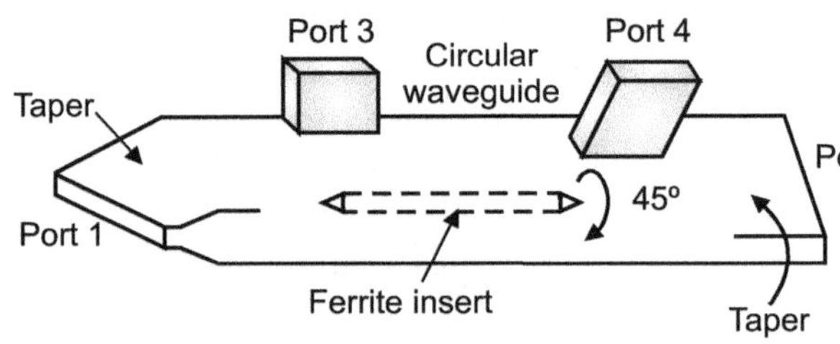

Fig. 3.19 : Faraday's rotation ferrite circulator

3.13.3 Working Principle

- When power is fed at port-1, it is converted to the TE_{11} mode in the circular waveguides. It passes by port-3 unaffected, because the electric field is not significantly cut, is rotated through 45° by the ferrite inside. When the wave travels along the magnetized ferrite, the direction of E field gets rotated by 45° to concide with the direction 'n'.
- The power incident at the port-1 appears at port-2. It does not get coupled to the port-4. Since, ports 2 and 4 are at 90° to each other. The power incident at the port-2 appears at the port-3 and the power incident at the port-3 appears at the port 4 and so on. It may be noted that ports-1 and 3 are on one end of the ferrite rod, whereas ports-2 and 4 are on the other end.
- It may be noted that this scheme cannot be easily adopted for a 3-port circulator. However, any 4 port circulator can be converted into a 3-port circulator by terminating one of the ports in a perfect short circuit. The most commonly used design of a 3 port circulator is however, based on a different principle. The circulators are used for isolation of transmitters and receivers connected to the same antenna as in Radar, or isolation of input and output terminals of amplifying device like a parametric amplifier.

3.13.4 Applications

1. It is used for isolation of transmitters and receivers connected to the same antenna of a Radar.
2. It is used for isolation of input-output terminals of a parametric amplifier.
3. It is used to couple transmitter to various receivers.

3.14 GYRATOR

3.14.1 Introduction

- A gyrator is a passive, linear, lossless, two-port electrical network element proposed in 1948 by *Bernard D. H. Tellegen* as a hypothetical fifth linear element after the resistor, capacitor, inductor and ideal transformer. Unlike four conventional elements, the gyrator is a non-reciprocal element.
- Gyrators permit network realizations of two or more port devices, which cannot be realized with just conventional four elements. In particular, gyrators make possible network realizations of isolators and circulators. Gyrators do not however change the range of one-port devices that can be realized.
- Although the gyrator was conceived as a fifth linear element, its adoption makes both the ideal transformer and either the capacitor or inductor redundant. Circuit that function as a gyrator can be built with transistors and op-amp using feedback. An important property of a gyrator is that it inverts the current-voltage characteristic of an electrical component of network.
- A gyrator can make a capacitor circuit behave inductively, a series LC circuit behave like a parallel LC circuit, and so on. It is primarily used in active filter design and miniaturization.

3.14.2 Definition

- **A gyrator is a passive, linear, lossless, two-port electrical element network element.**
- **A gyrator is a non-reciprocal ferrite two-port device that has a relative phase shift of 180° in forward direction and 0° phase shift in reverse direction.**
- *Bernard D. H. Tellegan* named the element *gyrator* as a *portmanteau of gyroscope* and the common device suffix *tor* (as in resistor, capacitor, transistor etc.)
- The *tor* ending is even more suggestive in Tellegen's native Dutch where the related element transformer is called *transformator*. The gyrator is related to the *gyroscope* by an analogy in its behavior.

3.14.3 Schematic Symbol

- The schematic symbol for a gyrator is as shown in Fig. 3.20.

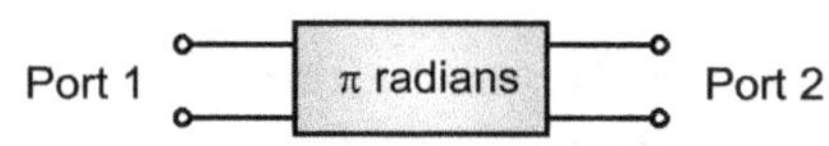

Fig. 3.20 : Symbol of a gyrator

3.14.4 Construction

- Gyrator consists of a circular to rectangular waveguide transition both at dominant mode. A twin circular ferrite rod tapered at both ends is located inside the circular waveguide surrounded by permanent magnets, which generate d.c. magnetic field for the operation of ferrite.
- A rectangular waveguide twisted by 90° is connected by input and reduces attenuation, and provide smooth attenuation of polarized wave.

3.14.5 Operation

- When signal is transmitted from port-1 to port-2, it offers phase shift of 180°, i.e. π radians and when signal is fed to port-2, it offers 0° phase shift to the signal. Hence, it is also known as *differential phase shift device.*

3.14.6 Applications

1. It can be used to invert the current-voltage characteristic of an electrical component or network.
2. It can be simulated as inductor and can be used to transform a load capacitance into an inductance.
3. It is especially useful in IC technology.
4. It is used to reduce the size and cost of a system by removing the need for bulky, heavy and expensive inductors.
5. It is extensively used in telephony devices that connect to a POTS system.
6. It is also used in most data access arrangements (DAA).
7. It is also widely used in Hi-Fi for graphic equalizers, parametric equalizers, discrete bandstop, and bandpass filters such as number filters and FM pilot tone filters.

3.15 COMPARISON OF ISOLATOR AND CIRCULATOR

Table 3.1

Circulator	Isolator
1. It has peculiar property that each terminal is connected to next clockwise terminals.	1. It uses the property of Faraday's rotation in the ferrite materials.
2. It has usually more than two ports.	2. It has only two ports.
3. It is a 4-port microwave device.	3. It is a 2-port microwave device.
4. It is used in parametric amplifiers, tunnel diode amplifiers and diplexers in Radars.	4. It is used in the output of an amplifier that is sensitive to the load conditions. It is also used in UHF or VHF transmitters and also in radio transmitters.

3.16 MICROWAVE POSTS

3.16.1 Brief History

- The exploitation of high frequency **radio waves** for heating substances was made possible by the development of **vacuum tube radio transmitters** around 1920. By 1930, the application of short waves to heat human tissues had developed into the medical theory of **dithermy.**
- At the 1933, **Chicago World's Fair, Westing house** demonstrated the cooking of foods between the metal plates attached to a 10 kW, 60 MHz **short wave transmitter**. The westing house team, led by I.F. Mouromtoseff, found that foods like steaks and potatoes could be cooked in minutes.
- In 1937, Bell laboratories of US has submitted an application for patent. However, lower frequency **dielectric hesting**, as described in the after mentioned patent is like **induction heating on electromagnetic** heating effect. This patent proposed radio frequency heating at 10 to 20 MHz (wavelength of 30 to 15 metre, respectively).
- Heating from microwaves that have a wavelength that is small relative to the cavity in a modern microwave oven is due to **"far field effects"** that are due to classical **electromagnetic radiation** that describes freely propagating light and microwaves suitably far from their sources. Nevertheless, the primary heating effect of all types of electromagnetic fields at both radio and microwave frequencies occur via the dielectric heating effect as polarized molecules are affected by a rapidly alternating electric field.

3.16.2 Definition

- A **microwave post** is commonly known as **microwave oven** or simply **microwave. A microwave post is an electric oven that heats and cooks food by exposing to electromagnetic radiations in the microwave frequency range.**
- Microwave ovens heat foods quickly and efficiently, because excitation is fairly uniform in the outer 25 to 38 mm, i.e. 1 to 1.5 inches, of high water content food items.

3.14.3 Construction

A microwave oven consists of the following components :

1. A high voltage power source.
2. High voltage capacitor.
3. A cavity magnetron.
4. A magnetron control circuit.
5. A short waveguide.
6. A metal cooking chamber.
7. A turn table.
8. A metal waveguide stirring fan.
9. A control panel.

3.16.4 Principle

- A microwave oven heats food by passing **microwave radiation** through it. Microwaves are a form of **ionizing** electromagnetic radiation with a **frequency** in the so called Microwave Region, i.e. 300 MHz to 300 GHz. Microwave ovens use frequencies in one of the ISM (Individual, Scientific, Medical) bands, which are otherwise used for communications amongst devices that don't need a license to operate, so they do not interfere with other vital radio services.
- Consumer ovens work around a nominal 2.45 GHz, a **wavelength** of 12.2 cm (4.80 inches) in the 2.4 GHz to 2.5 GHz ISM band, while large industrial/commercial ovens often use 915 MHz or 32.8 cm (9.15 inches).
- Water, fat, and other substances in the food absorb energy from the microwaves in a process called **dielectric heating**. Many molecules, such as those of water, are electric dipoles, meaning that they leave a partial positive charge at one end and a partial charge at the other, and therefore rotate as they try to align themselves with the alternating electric field of the microwaves.
- Rotating molecules hit other molecules and put them to motion thus dispersing energy. This energy, dispersed as molecular rotations, vibrations and/or translations in solids and liquids raise the temperature of the food, in a process similar to heat transfer by contact with a hotter body.

3.16.5 Hazards

1. High temperatures.
2. Metal objects.
3. Direct microwave exposure.
4. Chemical exposure.
5. Uneven heating.
6. Interference.

3.17 S.S. TUNERS

3.17.1 Definition

- **A tuner is a subsystem that receives radio frequency (RF) transmissions like radio broadcasts and converts the selected carrier frequency and its associated bandwidth into a fixed frequency that is suitable for further processing, usually because a lower frequency is used on the output.**
- The S.S. tuner is the shoot form **slide screw tuner.**

3.17.2 Key Features

1. Wide frequency range.
2. Large matching capability.
3. Low insertion loss.
4. Resetable design.
5. Low RF leakage.
6. Precision connectors.

3.17.3 Description

- The Masury series of coaxial slide screw (S.S.) tuners provide a convenient method of matching a wide range of impedances to a 50 ohm transmission line system. These tuners consist of a coaxial slot line section with a sliding carriage upon which two adjustable micrometers are mounted.
- The 0.001 inch reading micrometer controls a variable reactance element, whose magnitude is changed by varying its depth of insertion into the coaxial line and whose phase is varied by sliding the carriage along the line. Carriage travel is greater than one-half wavelength at the lowest operating frequency.
- A metric scale and vernier is provided to give precise readout of the carriage position to within 0.1 mm. Position locks are provided on both micrometer and the carriage.

3.17.4 Specifications

1. Frequency range.
2. Maximum correctable VSWR.
3. Travel at lowest.
4. Frequency.
5. Insertion loss.
6. Maximum VSWR with probe cut.
7. Nominal impedance.

3.17.5 Operating Principle

- Reflection on a solid-screw tuner is created by positioning the RF probes must be positioned very close to the central conduction, typically 0.05 to 1 mm away (2 to 40 thousands of an inch). The vertical position controls the amplitude and the horizontal position the phase of the rectification factor.
- Therefore, the most critical dimension in accurate tuner positioning is the vertical axis, whereas the horizontal reading is not very critical. The phasor error amounts to 2.4° per mm at 1 GHz. Focus tuners are supplied with a horizontal ruler that allow easy positioning within 0.1 mm or better.
- A digital horizontal movement reading feature is offered by same tuner. A fine micrometer screw provides vertical positioning. There again are differences. Focus uses bulkier and longer micrometer screws that move a non-rotating axis, whereas others use short screws and have a rotating link between micrometer screw and RF probe. This letter one is often cause of the backlash in the much more critical, vertical direction and inaccuracies the measurement.
- In any case, when repositioning a manual tuner, it is very important to come always from the some direction, in order to compensate for any backlash. RF reproducibility of manual focus tuners are remarkable.

3.17.6 Applications

1. Impedance matching in transmission lines.
2. Tuner.

3.18 ATTENUATORS

3.18.1 Definition

- **An attenuator is an electronic passive device that reduces the electrical power of a signal without appreciately distorting its waveform.**
- Attenuator is effectively opposite of an amplifier through the two work by different methods. While an amplifier provides gain, an attenuator provides less, or gain less than 1.

3.18.2 Characteristics

1. Nominal impedance, e.g. 50 Ω.
2. Frequency bandwidth, e.g. D.C. To 18 GHz.
3. Power dissipation.
4. Low standing wave ratio (SWR).
5. Accuracy.
6. Repeatability.
7. Flat frequency response.

3.18.3 Attenuator Circuits

1. n-type unbalanced.
2. n-type balanced.
3. T-type unbalanced.
4. T-type balanced.

- Basic circuits used in attenuators are pi pads (n-type) and T-pads (T-type). These may be required to be balanced or unbalanced networks depending on whether the line geometry with which they are to be used in balanced or unbalanced.

3.18.4 Types

1. Audio attenuators
2. RF attenuators.
3. Fixed RF attenuators.
4. Switched RF variable.
5. Resistor RF attenuators.
6. Variable attenuators.
7. PIN slide RF attenuators.
8. FET RF attenuators.

3.18.5 RF Attenuators

- Radio Frequency (RF) attenuators are typically coaxial in structure with precision connectors as ports and coaxial, micro strip or thin-film internal structure. Special waveguide structure as 5HF is required. The important characteristics are as mentioned above.
- The size and shape of the attenuator depend on its ability to dissipate power, RF attenuators are used as loads for and known as attenuation and protective dissipation of power in measuring RF signals.

3.18.6 Audio Attenuator

- A line-level attenuator in the preamplifier or a power attenuator after the power amplifier used electrical resistance to reduce the amplitude of the signal that reaches the speaker, reducing the volume of the output.
- A line-level attenuator has lower power handling, such as a half (0.5) watt potentiometer or voltage divider and controls preamplifier level signals, whereas a power attenuator has higher power handling capability, such as 10 watts or more, and is used between the power amplifier and the speaker.

3.18.7 Applications

- The application of RF attenuators are as under :
  1. Reduce signal gain.
  2. Improve impedance match.
  3. Variable voltage control.

3.19 PHASE SHIFTERS

3.19.1 Introduction

- A phase shifter is a control device found in many microwave communication, radar and measurement systems. In order to understand subtle difference between the two main generic types of phase shifter, both the true phase shifter and the true delay line must be defined.
- A true phase shifter can be defined as a control that has a flat group delay frequency response within its defined bandwidth of operation, the level of which does not change as the insertion phase is varied. The frequency phase characteristics of a true-phase shifter can be employed in multiple space diversity receiver combiners for aligning RF signal, with a pulse envelope without changing the timing of the pulse edges.
- However, they should not be employed in wideband beam-forming network for large aperture phased array antenna in order to avoid the effects of **phase squinting and pulse stretching.** A time shifter can be defined as a control device that has a flat group delay frequency.

3.19.2 Fundamental Concept

- Phase shifters are a critical component in many RF and microwave systems. Applications include controlling the relative phase of each element in a phasor array antenna in radar or steerable communications link and in cancellation loops used in high linearity amplifiers.

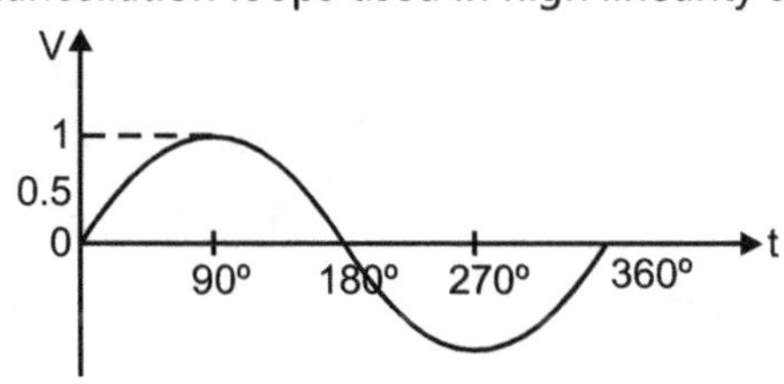

Fig. 3.21

- The phase of a single tone is meaningless until it is compared to another signal. If we look at a sine wave in the time domain, it will like a Fig. 3.21. If a second sine wave is added with a 90° phase offset from the first, it would like Fig. 3.22. If the phase of the second signal is varied or shifted, the sine wave is seen to move relative to the first sine wave.

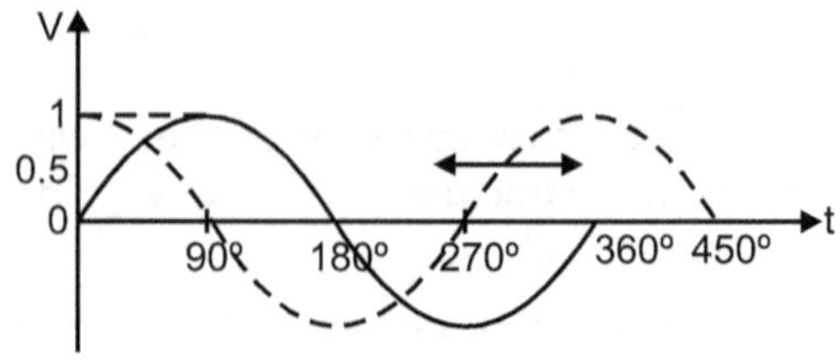

Fig. 3.22

- If the phase of the second signal is varied or shifted, the sine wave is seen to move relative to the first sine wave. Phase, therefore, is the fraction of a wavelength difference between two signals and can vary from 0° to 360°. This leads to an alternative way to view the relative phasor of a signal. If we think a single tone as having an amplitude and phase, then we can depict the signal on a polar diagram as in Fig. 3.23.

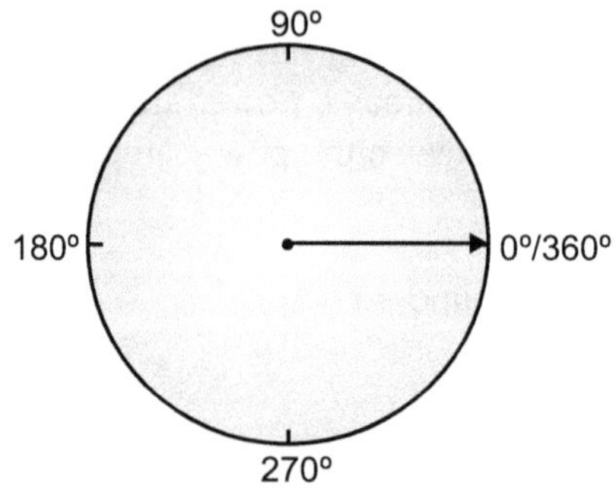

Fig. 3.23

3.19.3 Definition

- **Phase shift is a common term in the study of waveforms and communication of signals, which refers to the displacement of two signals, when propagated in a time domain.** This displacement can be caused by a signal processing device, such as an electronic amplifier or a low or high-pass filter that performs certain operations on the signal, causing the output signal phase to shift from its original signal phase. Phase shift is also known as phase change. The phase shift does not change the frequency of the signal.
- **A phase shifter is a control device that has a flat group delay frequency response within its defined bandwidth of operation, the level of which does not change as the insertion phase is varied.**

3.19.4 Types

- There are two types of phase shifters; **analog** and **digital**. Analog phase shifters provide a continuously variable phase to time delay. Digital phase shifters provide a discrete set of phase shifts or time delays. Discretization leads to quantization errors. Digital phase shifters require parallel bus control. The phase shifters are also of two types namely, **passive** and **active.**

3.19.5 Characteristics

1. Low insertion loss.
2. Equal amplitude for all phase stages.
3. Reciprocal networks.
4. True time delay.
5. Flat phase versus frequency response.

3.19.6 Advantages

(A) Analog Phase Shifters :

1. Lower loss
2. Lower cost of parts.
3. Sensitive to assembly variations.

(B) Digital Phase Shifters :

1. Immunity to noise on control lines.
2. More uniform performance.
3. Ability to achieve flat phase over wide bandwidth.
4. Less susceptible to phase pulling.
5. Easier to assemble.
6. Linearity.
7. Potentially higher power handling.

3.19.7 Applications

1. Frequency translators.
2. Phased arrays.
3. Phase errors.
4. Residual phase noise measurements.
5. Changing transmission phase angle.

3.20 CAVITY RESONATOR

3.20.1 Introduction

1. Musical instruments.
2. Radio transmitters.
3. Crystal oscillators.
4. Microwave receiver.

3.20.2 Definition

- **A resonator is a device or a system that exhibits resonance or resonant behaviour, i.e. it naturally oscillates with greater amplitude at same frequencies called resonant frequencies, than at other frequencies.**
- The oscillation in a resonator can be either **electromagnetic** or **mechanical**. A cavity resonator is an electronic device consisting of a space, usually enclosed by metallic walls, within which resonant electromagnetic fields may be excited and extracted for use in microwave systems.
- **A cavity resonator is one in which waves exist in a hollow space inside the device.**

3.20.3 Construction

- A cavity resonator is a hollow closed conductor such as a metal box as a cavity within a metal block, containing **electromagnetic waves**, i.e. radio waves, reflecting back and forth between the cavity's walls. When a source of radio waves at one of the cavity's **resonant frequencies** is applied, the oppositely moving waves from **standing waves** and the cavity ??? electromagnetic energy.
- Since, the cavity's lowest resonant frequency, the fundamental frequency, is that at which the width of the cavity is equal to a half wavelength, cavity is resonators are only used at microwave frequencies and above, where wavelengths are short enough that the cavity is conveniently small in size.
- Due to the reactance of their conductor walls, cavity reactors have very high **quality factor** (Q); that is their **bandwidth**, the range of the frequencies around the resonant frequency at which they will resonate, is very narrow. Thus, they can act as narrow **bandpass filters**. Cavity resonator are widely used as the frequency determining element in **microwave oscillators.** Their resonant frequency can be tuned by moving one of the walls of the cavity in or out, changing its size.

3.20.4 Types

1. Regulated cavity resonator.
2. Unregulated cavity resonator.
3. Coaxial cavity resonator.
4. Capacitive cavity resonator.
5. Inductive cavity resonator.
6. Waveguide cavity resonator.
7. Reentrant cavity resonator.

3.20.5 Operating Principle

- A physical system can have as many **resonant frequencies** as it has **degrees of freedom**; each degree of freedom can vibrate as a **harmonic oscillator.** Systems with two degrees of freedom, such as a mass on a spring, **pendulums** and **resonant transformers** can have two resonant frequencies.

- A **crystal lattice** composed of N atoms bound together can have N resonant frequencies. As the number of coupled harmonic oscillators grows, the time it takes to transfer energy from one to the next becomes significant. The vibrations in them begin to travel through the coupled harmonic oscillators in waves, from one oscillator to the next.
- The term **resonator** is often used for a homogeneous object in which vibrations travel as waves, at an approximately constant velocity, bouncing back and forth between the sides of the resonator. The material of the resonator, through which the waves flow, can be viewed as being made of millions of coupled moving parts, such as atoms.
- Therefore, they can have millions of resonant frequencies, although only a few may be used in practical resonators. The oppositely moving waves **interfere** with each other and at its **resonant frequencies** reinforce each other to create a pattern of standing wave in the resonator.
- The resonant frequencies of resonators called normal **modes** are equally spaced **multiples (harmonics)** of a lowest frequency called the **fundamental frequency**. If the medium inside the resonator is homogeneous then the waves travel at a constant speed, and that the shape of the resonator is rectilinear. If the resonator is homogeneous or has a non-rectilinear shape, like a circular **drumbead** or a cylindrical **microwave cavity**, the resonant frequencies may not occur at equally spaced multiples of the fundamental frequency. They are then called **overtones** instead of **harmonics.** These are several such series of resonant frequencies in a single resonator, corresponding to different modes of vibration.

3.21 MATCHED TERMINATION

3.21.1 Need

- In making measurement of waveguide components, it is desirable to absorb the power propagated down the waveguide. These are designed in such a way to absorb the maximum energy without having appreciable reflection assuring low VSWR. Therefore, there is a need of matched termination.

3.21.2 Definition

- A termination is a one part electronic device with an impedance that matches the characteristic impedance of a given transmission line.
- **A dummy load is a high power one port device intended to terminate a transmission line.** They are primarily employed to test high power microwave system at full power capacity. Low power coaxial loads are generally termed as **terminations** and typically handle one watt or less. Dummy loads can operate over full waveguide bands generally a more economical unit can be manufactured for use over narrower frequency ranges.
- **A terminator is a device, which is connected to the last peripheral device in a sequence or the last node in a network.**

3.21.3 Description

- A termination is attached to a certain terminal or post of a device to observe the power transmitted to that terminal or to establish a reference impedance at that terminal. Important parameters of a termination are its VSWR and power handling capacity. In a receiver, terminations are usually placed at various unconnected parts of components such as hybrid and power dividers to keep the VSWR of the signal path low.
- It is extremely important that the isolated part in a directional coupler and the unused port of a power divider, i.e. only three ports of a four way power divide are used, be properly terminated. All of the design considerations of directional couplers and power dividers are based on the fact that all ports will be reduced, which may severely degrade the performance of the receiver.
- The terminations are employed to terminate the unconnected ports on devices when measurements are being performed. They are useful as dummy antennas and as terminal loads for impedance measurements of transmission device, such as filters and attenuators.

- The resistive elements in most terminations are especially fabricated for use at microwave frequencies. Two types of terminations are employed as under :
  1. Resistance film elements.
  2. Molded resistive tapers.
- The **resistive film** is very thin compared to the skin depth and normally are very short relative to wavelength at the higher operating frequency. The **molded** taper consists of a dissipative material evenly dispersed in a properly cured dielectric medium. Both forms of resistive elements provide compact, rugged terminations suitable for most severe environmental conditions with laboratory stability and accuracy.
- Terminations should be properly matched to the characteristic impedance of a transmission line. The termination characteristics of primary concern are as under :
  1. Operating frequency range.
  2. Average power handling capability.
  3. Operating temperature range.
  4. VSWR.
  6. Size.
  7. Weight.
- Many microwave systems employ directional couplers, which require terminations on at least one port, and most have various modes of operation or test where terminations are needed on certain terminals. A matched termination of a generalized transmission line is ideally represented by an infinite length of that lines having small, but non-zero loss per unit length so that all incident energy is absorbed and none is reflected.
- Standard mismatches are useful as standards of reflection in calibrating reflectometer set-ups and other impedance measuring equipment. They are also used during testing to simulate specific mismatches, which would be encountered on the terminals of components once the component is installed in the circuit system.

3.21.4 Features

1. High performance.
2. Full absorbments of 50 and 75 Ω types.
3. Low VSWR.
4. Space saving design.
5. Light weight.
6. Usable over an extremely wide bandwidth.
7. Rolls compliant.

3.21.5 Applications

1. Networking equipments.
2. Optical transmission devices.
3. Network analyzers.
4. Bitter or ratio testing (BERTs) systems.
5. Fixed Wireless Access (FWA).
6. VSWR measurements (upto 65 GHz).
7. Absorb energy.
8. Preventing a signal.
9. Directional couplers.
10. Tee junctions.

Practice Questions

1. Define scattering parameters and their requirements.
2. Explain the analysis of microwave components.
3. Explain the need, definition and types of multiple junctions.
4. Explain construction and working principle of E-plane junction.
5. Explain construction and working principle of H-plane junction.
6. Explain construction and working principle of hybrid junction.
7. Explain principle, types and properties of directional couplers.
8. Explain construction and working principle of two hole directional coupler.
9. Explain the terms : Waveguide bands and corners.
10. Explain the term : Ferrite components.
11. Explain construction and working principle of isolators.
12. Explain concept, types and applications of circulation.
13. Explain construction and working principle of four part circulator.
14. Explain construction and working principle of gyrator.
15. Give comparison of isolator and circulator.
16. Explain construction and working principle of microwave posts.
17. Explain construction and working principle of slide screw tuners or S.S. tuners.
18. Give definition, characteristics and types of attenuators.
19. Explain construction and working principle of phase shifters.
20. Explain construction and working principle of cavity resonator.
21. Give need, definition, features and applications of matched termination.

✍ ✍ ✍

Chapter **4**

MICROWAVE MEASUREMENTS

Syllabus

Power measurements using Calorimeters and Bolometers, Measurement of Standing Wave Radio (SWR), Frequency and wavelength, Microwave bridges.

4.1 POWER MEASUREMENTS USIGN CALORIMETER

4.1.1 Introduction

- Electricity is measured in units of power called Watts.
- A watt is the unit of electrical power equal to one ampere under the pressure of one volt. One watt is a small amount of power. Some devices require only few watts to operate, and other devices require larger amounts. The power consumption of small devices is usually measured in watts, and the power consumption of larger devices is measured in kilowatts (kW).
- Electricity generation capacity is often measured in multiples of kilowatts, such as mega watts (MW) and giga watts (GW), and one GW is 1000 MW (or 10^9 watts). A watthour (Wh) is equal to the energy of one watt steadily supplied to, or taken from, an electric circuit for one hour. The amount of electricity that a power plant generates or an electricity utility customer uses is typically measured in kilowatt hour (kWh). One kWh is one kilowatt generated or consumed for one hour.
- Electric utilities measure the electricity consumption of their customers with meters that are usually located on the outside of the customer's property, where the power line enters the property. Now many utilities use electronic smart (digital) meters, which provide wireless access to the meter's power using data, to measure electricity consumption in real-time.
- The measuring of power consumption of a device or the dissipated power of a component is often a source of errors. When the power is measured, it is necessary that both the voltage and current are measured at the same time, and that the relationship between these two is taken into account. The ways the voltage and current are measured have a great influence on the accuracy.

4.1.2 Important Terms

1. **Real Power (P) :** The real power is the only correct parameter for measuring power. The real power is what is actually converted into heat or work. The real power multiple measurements may always be added together. The real power P is calculated by averaging the products of the instantaneous values of voltage V and current I.
2. **Apparent Power (S) :** The apparent power is calculated from the effective voltage and current. It may never be added together to calculate. The total apparent power, which must always be measured.
3. **Reactive Power (Q) :** The reactive power is also called wattless power. It is calculated as follows :

$$Q = V_{eff} \cdot I_{eff} \cdot \sin\theta \qquad \dots (4.1)$$

To calculate the reactive power, an additional phase measurement has to be carried out. It applies only to pure sine shaped voltages and currents. The measurement is not valid, if there is a non-linear load.

4.1.3 Measuring Arrangement

- Fig. 4.1 shows a widely used measuring arrangement with an individual volt and ampere meter. The applications are very restricted. In general, this configuration is conveniently applied for D.C., because the voltages will normally be reasonably constant.
- If this arrangement is used for A.C. applications and the voltage and current shapes are not sinusoidal, then the instruments must measure the true RMS values.

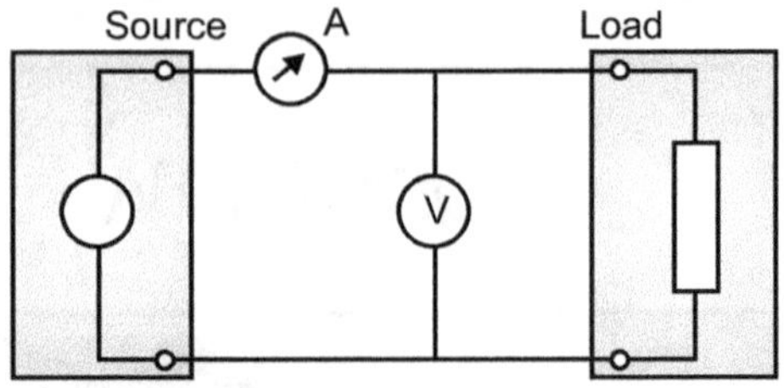

Fig. 4.1 : Power measurement with volt and ampere meters

4.2 POWER MEASUREMENT USING CALORIMETER AND BOLOMETER

4.2.1 Introduction

- In the field of microwave engineering, there arises numerous applications. Therefore, while using dissimilar applications, we frequently come across the need of measuring different values such as power, attenuation, phase shift, VSWR, impedance, etc. for the effective usage.

4.2.2 Measurement of Power

- The microwave power measurement is the average power at any position in waveguide. The power measurement can be of following three ways :
  1. Measurement of low power.
  2. Measurement of medium power.
  3. Measurement of high power.

4.2.3 Measurement of Low Power by Bolometer

- The measurement of microwave power around 0.01 mW to 10 mW can be assumed as the measurement of low power. A device, which is used for low microwave power measurements is called **Bolometer.** Thus, **Balometric technique** is used for measurement of low microwave power.
- The element used in Bolometer could be of positive or negative temperature coefficient. For instance, a Bolometer has a positive temperature coefficient, whose resistance increases with the increase in temperature. Thermistor has negative temperature coefficient, whose resistance losses with the increase in temperature.
- Any of them can be used in Bolometer, but the change in resistance is proportional to microwave power applied for measurement.
- This bolometer is used in a link of the arms as one so that any inequality caused affects the output. A usual instance of a bridge circuit using a Bolometer is as shown in Fig. 4.2.
- The mA here provides the value of the current passing. The battery is adaptable, which is mixed to obtain balance, when an inequality is caused by the behaviour of the Bolometer. This change which is made in D.C. battery voltage is relative to the microwave power. The power usage capacity of this circuit is limited.

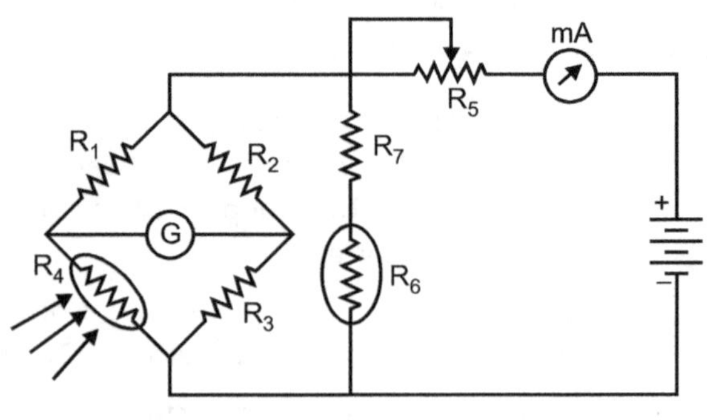

Fig. 4.2 : Power measurement by Bolometric technique

4.2.4 Measurement of Medium Power by Colorimeter

- For the measurement of microwave power around 10 mW to 1 W, we can be agreed as the measurement of medium power. A special kind of load is engaged, which typically upholds a certain value of specific heat. The power to be measured is practical at its input, which proportionally changes the output temperature of the load that it previously maintains.
- The difference in temperature rise identifies the input microwave power to be the load. The bridge balance method is now used to get the output. The heat transfer method is used for the measurement of power, which is a **colorimetric technique.**

4.2.5 Measurement of High Power

- The measurement of microwave power around 10 W to 50 kW, can be agreed as the measurement of high power. Colorimetric wattmeters, which can be of dry and flow type. The dry type is called so as it uses a coaxial cable, which is filled with dielectric of high hysterisis loss, whereas the flow type is named so as it uses water or oil or some liquid, which is a good absorber of microwaves.
- The alternation in temperature of the liquid earlier and after entering the load, is taken for the calibration of values. The limits in this method are like flow determination, calibration and thermal inertia, etc.

4.3 MEASUREMENT OF STANDING WAVE RATIO (SWR)

4.3.1 Introduction

- SWR is usually measured using a dedicated instrument called an **SWR meter.** Since, SWR is a measure of the load impedance relative to the characteristic impedance of the transmission line in use, which together determine the reflection **coefficient.** A given SWR meter can only interpret impedance, it sees in terms of SWR, if it has been designed for that particular characteristic impedance.
- In practice, most transmission lines used in these applications are **co-axial cables** with an impedance of either 50 or 15 ohms, so most SWR meters correspond to one of these. Checking the SWR is a standard procedure in a radio station. Although the same information could be obtained by measuring the load's impedance with an **impedance analyzer** or **impedance bridge.**
- **The SWR meter** is simple and most robust for this purpose by measuring the magnitude of the impedance mismatch at the transmitter output, it reveals problems due to either the antenna or the transmission line.

4.3.2 Definition

- **In radio engineering and telecommunications, standing wave radio (SWR) is a measure of impedance matching of loads to the characteristic impedance of a transmission line or waveguide.**
- Impedance mismatches result in standing waves along the transmission line, and **SWR is defined as the ratio of the partial standing wave's amplitude at an antinode (maximum) to the amplitude at a node (minimum) along the line.**
- **The standing wave ratio (SWR) is usually thought of in terms of the minimum and maximum A.C. voltages along the transmission line. This is called the voltage standing wave ratio (VSWR).** For example, the VSWR value 1.2 : 1 denotes the A.C. voltage due to standing wave along the transmission line reaching a peak value 1.2 times that of the minimum A.C. voltage along that line.
- **The SWR** can be well defined as the **ratio of the maximum amplitude to minimum amplitude of the transmission line's currents, electric field strength, or the magnetic field strength. Neglecting transmission loss, these ratios are identical.**
- **The power standing wave ratio (PSWR) is defined as the square of the VSWR.** However this depreciated terminology has no physical relation to actual powers involved in transmission.

4.3.3 Impedance Matching

- SWR is used as a measure of **impedance matching** of a load to the **characteristic impedance** of a transmission line carrying **radio frequency (RF)** signals. This especially applies to transmission lines connecting **radio transmitters** and receivers with their **antennas,** as well as similar uses of RF cables such as **cable television** connections of TV receivers and **distribution amplifiers.**
- Impedance matching is achieved, when the source impedance is the complex conjugate of the load impedance. The easiest way of achieving this and the way that minimizes losses along the transmission line, is for the imaginary part of the complex impedance of both the source and load to be zero, that is pure resistances equal to the characteristic impedance of the transmission line.
- When there is a **mismatch** between the load impedance and the transmission line, part of the forward sent towards the load is reflected back along the transmission line towards the source. The source then sees a different impedance than it expects, which can lead to lesser or in some cases, more power being supplied by it, the result being very sensitive to the **electrical length** of the transmission line.
- Such a mismatch is usually undesired and results in standing waves along the transmission line, which magnifies transmission line losses, i.e. significant at higher frequencies and longer cables. The SWR is a measure of the depth of those standing waves and is there a measure of the matching of the load to the transmission line.
- A matched load would result in a SWR of 1 : 2 implying no reflected wave. An infinite SWR represents complete reflection by a load unable to absorb electrical power, with all incident power reflected back towards the source.

4.3.4 Methods for Measuring SWR

- Many different methods can be used to measure standing wave ratio (SWR). The most intuitive method uses a slotted line, which is selection of transmission line with an open slot, which allows a probe to detect the actual voltage at various points along the line. Thus the maximum and minimum values can be compared directly.
- This method is used at VHF and higher frequencies. At lower frequencies, such lines are impractically long. Directional couplers can be used at high frequency (HF) through microwave frequencies. Other types of directional couplers sample the current and voltage at a single point in the transmission path and mathematically combine them in such a way as to represent the power flowing in one direction.
- The common type of SWR/power meter used in amateur operation may contain a dual directional coupler. Other types use a sample which can be rotated 180° to sample power flowing in either direction. Unidirectional couplers of this type are available for many frequency ranges and power levels and with appropriate coupling values for the analog meter used.
- The forward and reflected power measured by directional couplers can be used to calculate SWR. The computations can be done mathematically in analog or digital form or by using graphical methods built into the meter as in additional scale or by reading from the crossing point between two needless on the same meter.
- The above measuring instruments can be used in line, that is, the full power of the transmitter can pass through the measuring device so as to allow continuous monitoring of SWR. Other instruments, such as network analyzers, low power directional couplers and antenna bridges use low power for the measurement and must be connected in place of the transmitter.
- Bridge circuits can be used directly to measure the real and the imaginary parts of load impedance and to use these values to derive SWR. These methods can provide more information than just SWR or forward and reflected power. Stand alone antenna analysers use various measuring methods and can display SWR and other parameters plotted against frequency.
- By using directional couplers and a bridge in combination, it is possible to make in-line instrument that reach directly in complex impedance or in SWR. Stand alone **antenna analyzers** are also available that measure multiple parameters.

4.4 MEASUREMENT OF FREQUENCY

4.4.1 Introduction

- Frequency measurement is an important issue in Electrical Engineering. Electric power systems have become increasing complex over the last decade. The use of distributed generation, connection of non-linear loads and presence of unexpected system faults are the main causes of frequency variations. Power quality parameter includes frequency as an important index for measurement of power system stability.
- In power system protection and control schemes, it is necessary to accurately measure and track the fundamental power frequency under sinusoidal and non-sinusoidal conditions. The frequency deviation is a good indicator of the system abnormal operating conditions. Measurement of frequency is fast and accurate, if the signal is purely sinusoidal.
- However, in reality the measured signals are distorted by higher harmonics and have time varying parameters. Since, power electronics technology is increasingly applied in the power systems, it is required to detect frequency more precisely and faster in the presence of harmonics and noise. So, many estimation methods have been developed offering different approaches to the problems, such as zero crossing methods, FET leakage stationary phasor phase changes, **Kalman filtering** and least squares fitting. The speed and accuracy of frequency measurement varies in each method. It is likely that method giving fast and accurate frequency measurement under non-sinusoidal conditions will be of great help to the utility.

4.4.2 Definition

- **Frequency is the number of occurrences of a repeating event per unit time.** It is also referred to as **temporal frequency,** which emphasizes the contrast to **spatial frequency** and **angular frequency.**
- **The period is the duration of time of one cycle in a repeating event, so the period is the reciprocal of the frequency.** Frequency is an important parameter used in science and engineering to specify the rate of oscillatory and vibration phenomenon, such as mechanical vibrations, audio signal (sound), radio waves and lights.
- The relation between the frequency f and the period T, of a repeating event or oscillation is given by,

$$f = \frac{1}{T} \quad \dots (4.3)$$

4.4.3 Basic Principle

- Frequency measurement is very important application of both counting and timing. Fundamentally, frequency measurement is a measure of how many times something happens within a certain known period. The use can be as diverse as how many counts are received per minute in a Geiger count, how many cycles per second, i.e. Hertz (Hz), there are in an electronic or acoustic measurement, or how many wheel resolutions, there are per unit of time in a speed measurement.
- Both a counter and a timer are needed, the timer to measure the reference time and the counter to count the number of events within that time.

4.4.4 Frequency Tracking Methods

1. Zero crossing methods.
2. Quadratic form.
3. Discrete Fourier transform with phase compensation.
4. Demodulation.
5. Decomposition of single phase into orthogonal components.
6. Linear estimation of phase.
7. Non-linear least squares estimation.

4.4.5 Measurement Procedure

- The proposed algorithm is a modified version of Sezis method. This algorithm uses phase angle and amplitude information in order to compute frequency. The main advantage of this method is that it avoids waiting between zero crossings of input signal.
- The block diagram for measurement of frequency using digital filters is shown in Fig. 4.3.

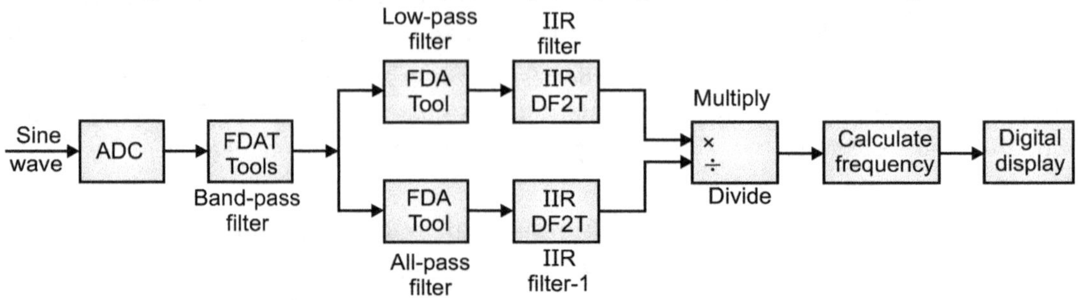

Fig. 4.3 : Black diagram of frequency measurement of single-phase system

- The input signal used for measurement of frequency is a single-phase voltage signal. Single voltage signal is available anytime irrespective of the load. Therefore, a voltage signal is used in the model. The time needed to reach the steady state is over three cycles, but it can be reduced, if filter with shorter order is used.
- The digital filters are initially designed and simulated on MATLAB. The magnitude response plot of various filters used in this block diagram are studied on MATLAB and depending upon their performance under various sinusoidal and non-sinusoidal conditions varying trails were performed to finalise the order and type of filters.
- The input signal is sampled by the ADC block which converts the analog voltage signal into digital sample values, which are further filter by band pass filter, which is a 40 order FIR (Finite Impulse Response) filter used with a pass-band frequency range from 30 to 100 Hz. The time required to reach steady state is 2 cycles due to 40 order filter increasing the order will increase the time to reach state.
- The other two filters used in blocks are FIR filters, there are mainly low-pass and all-pass filters. Both the filters are digitally designed in such a way that the phase angle change to the input signal is same for both the cases. The digital instantaneous values of the voltage signal are processed through band-pass and then through all-pass and low-pass filters, whose design and the response of filters were explained. After passing the instantaneous values through the low-pass and all-pass filters, the instantaneous values are then processed through an IIR filter, which is designed using equation (4.1). The design of IIR filter is not covered is Sezis model. The IIR filter is required to obtain a value proportional to signal amplitude.

$$y_c[n] = y_c[n-1] + 0.1\,(x_c^2[n] - x_c^2[n-40]) \qquad \dots (4.2)$$

where, y_c is IIR filter output and x_c is input filter data values.

4.5 MEASUREMENT OF WAVELENGTH

4.5.1 Introduction

- Assume a sinusoidal wave moving at a fixed wave speed. Wavelength is inversely proportional to frequency of wave. Waves with higher frequencies have shorter wavelengths and lower frequencies have longer wavelengths. Wavelength depends on the medium, for example, vacuum, air, or water that a wave travels through.
- Examples of waves are sound waves, light, water waves and periodic electrical signals in a **conductor**. A **sound** wave is a variation in air, pressure, while in **light** and other electromagnetic **radiation**, the strength of the electric and the **magnetic** fields vary. Water waves are variations in the height of a body of water. In a crystal **lattice vibration** atomic positions vary.

- **The range of wavelengths or frequencies for wave phenomena is called a spectrum.** The name originated with the **visible light spectrum**, but now can be applied to the entire **electromagnetic spectrum** as well as to a **sound spectrum** or **vibration spectrum.**
- In a linear media, any wave pattern can be described in terms of the independent propagation of sinusoidal components. The wavelength λ of a sinusoidal waveform travelling at constant speed v is given by,

$$\lambda = \frac{v}{f} \qquad \dots (5.4)$$

where, v is the phase speed of the wave and f is the frequency of the wave.

- In a **dispersive media**, the **phase speed** is **phase velocity** itself depends upon the frequency of the wave, making the **relationship between wavelength and frequency** non-linear.

4.5.2 Definition

- Apart from amplitude, another important measure of wave size is the wavelength. **Wavelength is the distance between two corresponding points of the same phase on adjacent waves.** It can be measured as the distance between two adjacent crests of a **transverse wave** or two compressions of a **longitudinal wave**.
- Wavelength is usually measured in meters. It is related to the energy of a wave. Short wavelength waves have more energy than long wavelength waves of the same amplitude.
- **The inverse of the wavelength is called spatial frequency.** Wavelength is commonly designated by the Greek letter lamda (λ). The term wavelength is also sometimes applied to **modulated** waves and to the sinusoidal envelopes of modulated waves formed by interference of several sinusoids.

4.5.3 Procedure of Measurement

- Set-up the 12 V, 36 W line filament lamp high at one end of the laboratory, so that you can see it clearly. Place a green filter infront of the lamp. If necessary, increase that applied voltage to 14 to 15 V so there is enough light.
- Hold a metre rule straight out infront of you towards the lamp, with the near end of the rule at your face. Hold the diffraction grating against the near end of the meter rule and look at the lamp through it.
- Ask your partner to place another meter rule at 90° to your metre rule and move it along unitil you see it in the green region of your bright spectrum. Note the distance x along your partner's ruler from the pencil to the far end of your ruler.
- When you have made your observation, record it and change places with your partner so that he or she can take their turn.
- Divide your measurements x by the length of your ruler, 100 cm. This gives you tan A, where A is the angle between the line of direct while light and the light to the green in the spectrum marked by the pencil. From tan A, use your calculation to find the angle A, and from this find sine A.
- Use the formula d sine A equal to wavelength to calculate the wavelength of green light. You will need the value of d, spacing, i.e. distance from one ruling to the next. If the grating has 300 lines per mm, then the spacing is $\frac{1}{3 \times 10^5}$ m.

4.6 MICROWAVE BRIDGES

4.6.1 Definition

- **A microwave bridge is a microwave circuit equivalent to an ordinary electrical bridge used to measure impedance, consisting of six waveguide sections arranged to form a multiple junction.**

4.6.2 Methods of Measurement

- More accurate methods of measuring microwave attenuation and phase are constantly being sought, particularly for such applications as **plasma diagnostics.** The microwave bridge technique described here was developed for the study of a **quiescent plasma**, having an electron density of 1015 to 1018 ma corresponding to a plasma frequency of 1010 to 1011 sa.
- The plasma has a broad dimension of OA 0.3 m. For such a plasma had a broad dimension of OA 0.3 m. For such a plasma a probing frequency of 10 GHz was considered to be the most suitable, at this frequency the attenuation and phase shift.
- A balanced microwave bridge was used to measure the absorption. 5 mm power was obtained from the second harmonic generated in a non-linear silicon crystal driven by a 1 cm reflex Klystron oscillator.
- The 5 mm radiation was then divided between two nearly identical wave guides and recombination in a bridge T-junction. One output arm of the T-fed a matched load and the other fed a **harmonic converter.** The converter mixed the incoming 5 mm radiation with the second harmonics of another 1 cm local oscillator, and the heterodyne signal current was fed to 24 MHz amplifier.
- Provisions were made for inserting a step-attenuator pad in the amplifier cascade after 3 stages of amplification. Both the signal oscillator and local oscillator were frequency stabilized to a high mode of a 3 cm cavity.
- The frequency of the signal generator fundamental was measured with the M.I.T. frequency standards.
- The signal generator fundamental was amplitude modulated with a rotating attenuator at 30 Hz. The detected output from the receiver could then be filtered for 30 cycle modulations. This modulation was amplified and converted to D.C. with a phase detector. The signal was read on a milliammeter and the 30 cycle modulations mere by provided a convenient monitoring current to amplify after filtering.
- In operation, the sample arm of the bridge was filled with tank oxygen at 80 cm - Hg, and the bridge balanced for no receiver output by means of the RF phase shifter and attenuator in the sample arm. The sample was then pumped out and the bridge rebalanced as well as possible by slowly admitting tank argon into the sample arm to an appropriate pressure.
- The final minimum was achieved with readjustment of the RF phase shifter. The use of argon for rebalancing allowed the bridge to work under constant impedance conditions. This procedure proved necessary for accuracy. Argon was chosen because its dielectric constant is similar to that of oxygen. Since argon is monotomic, it has no absorption in the region of these measurements.

Practice Questions

1. Explain the following terms : (i) Real power, (ii) Apparent power, (iii) Reactive power.
2. Explain measuring arrangement with suitable diagram.
3. Explain measurement of low power by bolometer.
4. Explain measurement of medium power by colorimeter.
5. Explain measurement of high power.
6. Define the term standing wave ratio.
7. Define the term impedance matching.
8. Explain measurement of standing wave ratio.
9. Define the term frequency.
10. Explain frequency measurement by single-phase system.
11. Define the term wavelength.
12. Explain procedure of measurement of wavelength.
13. Define the term microwave bridge.
14. Explain method of measurement of microwave bridge.

✍✍✍

www.ingramcontent.com/pod-product-compliance
Lightning Source LLC
LaVergne TN
LVHW062209171026
850499LV00044B/1896
* 9 7 8 9 3 8 9 9 4 4 3 3 4 *